Leading processes to lead companies:
Lean Six Sigma

Gabriele Arcidiacono · Claudio Calabrese · Kai Yang

# Leading processes to lead companies:
# Lean Six Sigma
## Kaizen Leader & Green Belt Handbook

 Springer

**Gabriele Arcidiacono**
Marconi University of Rome (Italy)

**Claudio Calabrese**
Lean Six Sigma Expert

**Kai Yang**
Wayne State University, Detroit (USA)

ISBN 978-88-470-2347-5       ISBN 978-88-470-2492-2 (eBook)
DOI 10.1007/ 978-88-470-2492-2

Library of Congress Control Number: 2011931733

Springer Milan Dordrecht Heidelberg London NewYork

Cover-Design: Simona Colombo, Milan
Typesetting: PTP-Berlin, Protago T$_E$X-Production GmbH, Germany (www.ptp-berlin.eu)
Printing and Binding: Grafiche Porpora, Segrate (MI)

Springer-Verlag Italia s.r.l., Via Decembrio 28, I-20137 Milano
Springer is a part of Springer Science+Business Media (springer.com)

# Foreword

How and why should you become "Green Belt"?

It is not easy to explain *how* with just a foreword, but this book - which is very *lean* and well done - certainly represents the ideal tool to find the right answer to this question.

In my opinion, however, the book is even more effective for those people who are asking themselves "why should I become Green Belt?" I may try to give my personal contribution to this second question.

One possible answer may be: *because my Company has decided that I have to, in order to embark on process of in-house business improvements.*

# Foreword

Apparently this is the least appropriate answer, but it is not true. As a matter of fact, the value you acquire by entering into the training mechanism, by managing an improvement project following the proposed methodology, completing it and, finally, getting the Green Belt Certificate, makes appropriate even this answer.

However, we may provide several answers to the fatidic question and we may consider valid each of them, but most probably, the most appropriate one shall be *because I believe in the Lean Six Sigma methodology and in my professional growth!*

When you work with three inseparable elements like Products, Processes, People, we may surely consider the Process as the one which may positively influence the other two.

# Foreword

Better and suitable processes help people professional growth, complete their skills profile and support them in providing better products and services.

Nearby, however, I have my own and personal answer which I believe may complete the one I have already considered as the most appropriate: *I've become Green Belt because I have so strongly believed in this methodology to persuade my company to join it. I've found out both efficiency and effectiveness of the method, but above all, I've deeply needed it. I was needing to wholly lead the processes and to speak the same language of a team of people and, therefore, of a whole company which began to run at higher speed: the results were right there.*

# Foreword

And so, thanks to the authors because, coming at the end of this handbook, I am sure that the reader, the manager and the professional will find an obvious and easy answer to a simple question: Why have I become a Green Belt?

Rome, October 2011

*Massimo Scaccabarozzi*
(Certified Green Belt, Chairman of Farmindustria,
CEO of Janssen Pharmaceutical Company of Johnson & Johnson)

# Preface

This Minibook is a brief guide for Green Belt during a Lean Six Sigma project management or for Kaizen Leader during a process improvement activity. Having theoretical concepts and practical examples, it is a handbook for a quick consultancy.

The authors' idea comes from companies' needs to analyze useful information and get to know different kinds of processes in depth.

The set of Six Sigma tools are explained through Minitab 16, the latest release of the most widely used statistical software.

*The Authors*

# Table of contents

# Table of contents

# Table of contents

# Table of contents

LEAN SIX SIGMA MINIBOOK

XIV

# Table of contents

# Introduction: Six Sigma

*Six Sigma* (G) is a proven business strategy (structured according to the DMAIC phases) to measure, analyze and improve the performance in terms of operational excellence.

The methodology, thanks to a wide range of qualitative and quantitative tools, aims to optimize the manufacturing and transactional processes through reduction of their variability.

The 5 stages in the DMAIC approach are:

- DEFINE
- MEASURE
- ANALYZE
- IMPROVE
- CONTROL

(G) *This symbol indicates that the word is included in the Glossary (page 325)*

Arcidiacono G., Calabrese C., Yang K.: Leading processes to lead companies: Lean Six Sigma.
DOI 10-1007/978-88-470-2492-2, © Springer-Verlag Italia 2012

# Introduction: DMAIC approach

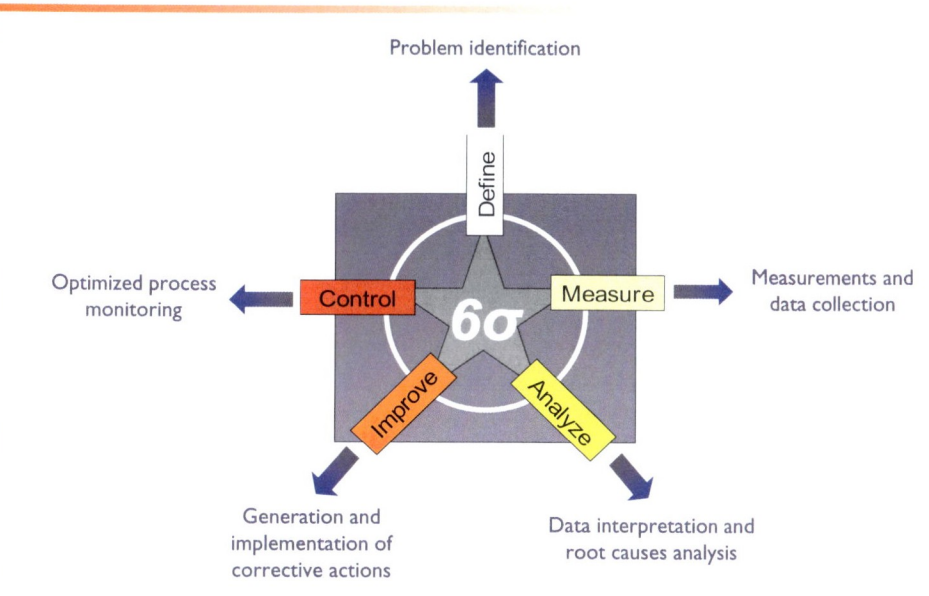

# Introduction: *Lean methodology*

*Lean* (G) methodology aims to relentlessly identify and eliminate waste in order to maximize speed and flexibility of business processes in order to deliver <u>what</u> is needed, <u>when</u> needed and in the <u>quantity</u> needed by the Customer.

Terms like *Lean Manufacturing* or *Lean Production* are deliberately <u>not</u> used, as the *Lean* method can be widely used in a variety of processes such as production processes and transactional processes, for example:

- *Lean Production* or *Manufacturing* for production processes
- *Lean Office* for service/support processes
- *Lean Design* inside the Research & Development process

# Introduction to "Waste"

What is the meaning of "**waste**"(G)?

It is the use of resources (time, material, labor, etc.) for doing something that customers are not willing to pay for, and so it does not add value to the product or service provided.

Eliminating waste improves the value of products and services.

*The Lean* "philosophy" highlights 8 macro-categories of waste:

- Overproduction
- Defects
- Transportation
- Inventory
- Waiting
- Over-processing
- Motion
- Underutilized people

LEAN SIX SIGMA MINIBOOK

# The 8 Wastes

| Waste Category | Description | Root Causes | Goals |
|---|---|---|---|
| Overproduction | Overproduction happens when a process produces more products/services than necessary | • Batch Production<br>• Production on forecast | Produce just the necessary, in the right time at the right quality |
| Defects | Production of defective parts/services that can't be sold to the Customer. Defects can be scraps or reworks, which add tremendous costs to organizations | • Lack of standardization<br>• Lack of training<br>• Lack of error proofing system<br>• Poor quality of supply<br>• Obsolete process | Produce "right first time"<br>Stop the process when the defect occurs, solve the problem in order to remove it definitively |
| Transportation | Unnecessary transport of materials | • Batch Production<br>• Inefficient layout<br>• Long set-up time | Minimize the movement by arranging processes in close proximity to each other |
| Inventory | Too many finished goods in inventory, WIP inventory, raw material inventory | • Batch Production<br>• Long set-up time<br>• Bottleneck<br>• Lack of continuous flow<br>• Push organization | The inventory must be dimensioned based on the real actual usage and on the supplier delivery time |
| Waiting | Customer waiting, waiting for materials, waiting for employees | • Bottleneck<br>• Lack of continuous flow<br>• Lack of standardization<br>• Unbalanced workload | Maximize "value adding" time to reduce waiting and to arrange processes in a continuous flow approach |
| Over-processing | Unnecessary processes or operations | • Not Value Added activity<br>• Lack of investigation of Customer needs<br>• Activity by "tradition" | Optimize Value Added activities to remove all the unnecessary steps |
| Motion | Not Value Added movement of people and machines | • Inefficient layout and process flow<br>• Lack of standardization | Remove unnecessary motion and improve disposition of material in the workplace |
| Underutilized People | Not using people's skills, people are seen as a source of labor and are not involved in finding solutions/opportunities to improve processes | • Lack of involvement<br>• Old Culture | People are the most important resource in a company: let's involve them as much as possible in company activities |

# Introduction: *Lean Thinking approach*

Every time a Lean expert looks at a process optimization he/she must consider the 5 *Lean Thinking* principles:

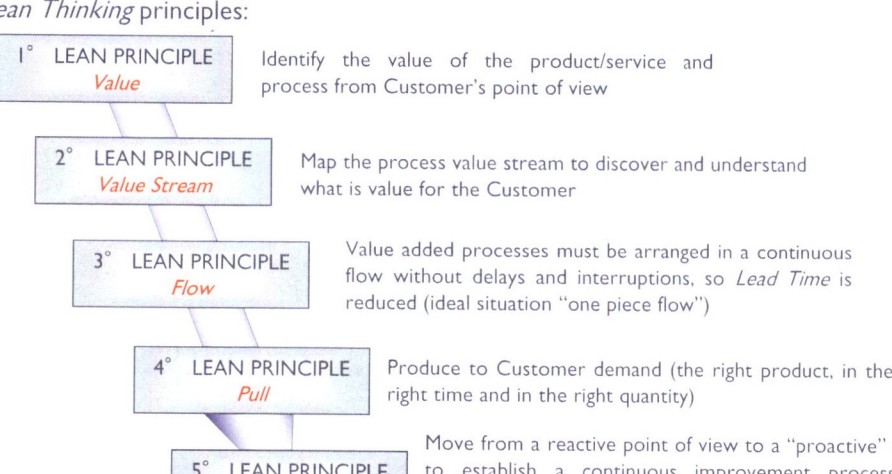

1° **LEAN PRINCIPLE**
*Value*

Identify the value of the product/service and process from Customer's point of view

2° **LEAN PRINCIPLE**
*Value Stream*

Map the process value stream to discover and understand what is value for the Customer

3° **LEAN PRINCIPLE**
*Flow*

Value added processes must be arranged in a continuous flow without delays and interruptions, so *Lead Time* is reduced (ideal situation "one piece flow")

4° **LEAN PRINCIPLE**
*Pull*

Produce to Customer demand (the right product, in the right time and in the right quantity)

5° **LEAN PRINCIPLE**
*Perfection*

Move from a reactive point of view to a "proactive" one, to establish a continuous improvement process of performance (looking for new Customer expectations and new possibilities to eliminate waste)

# Introduction: *Kaizen Event approach*

## Objective:

- KAIZEN[G] means "to become good through change". A Kaizen event is a focused effort to improve activity. A multi-functional team is created and, for a period of 3-5 days, focuses on resolving a specific problem. The main objectives of a Kaizen event are to solve problems and/or eliminate "waste"

## When to use it:

- Kaizen events are mostly used to develop activity such as:
  - 5S
  - Standard Work
  - Changeover Reduction (SMED)
  - Kanban implementation
  - Flow Improvement
  - Cell Design
  - Problem solving activities
  - TPM & OEE implementation

# Introduction: *Kaizen Event approach*

## 1. BEFORE KAIZEN EVENTS

| IDENTIFY OPPORTUNITY | SET THE TARGET | KAIZEN TEAM |
|---|---|---|
| The activity should be linked to one or more company goals | Identify the goal of the event; for example:<br>- Space reduction<br>- Inventory reduction<br>- Cycle time reduction<br>- Productivity<br>- Customer Satisfaction | Identify the team (typically 5-8 people); for example:<br>- Kaizen Leader and Co-Leader<br>- Manufacturing Engineer<br>- Product Engineer<br>- Material Planner<br>- Safety<br>- Quality Specialist |

## 2. DURING KAIZEN EVENTS (3-5 days)

| INITIAL TRAINING | CURRENT STATE ANALYSIS | FUTURE STATE ANALYSIS | RUN THE NEW PROCESS | ACTION PLAN & CLOSURE |
|---|---|---|---|---|
| The initial training aims to introduce the activity for the next 3-5 days and must not be longer than 2 hours | Data collection, problem description, Process Mapping, Current State performance definition | Identify improvements, reduce or eliminate waste, describe Future State, implement actions | Test the action implemented, new data collection and results analysis | Action plan of open points to be implemented (Kaizen Newspaper), prepare presentation for Kaizen Event, closure activity |

## 3. AFTER KAIZEN EVENTS

| ACTION PLAN IMPLEMENTATION | NEW PROCESS CONSOLIDATION | STANDARDIZATION OF NEW PROCESS | EXTENTION OF THE RESULTS IN NEW AREAS |
|---|---|---|---|
| Kaizen Newspaper implementation | Standardize process measurements | Control plan of process measurements | Identify opportunity to deploy the results on similar areas |

LEAN SIX SIGMA MINIBOOK

# Introduction: Kaizen Events vs LSS Project

| Elements of comparison | Lean Six Sigma Project | Kaizen Events |
|---|---|---|
| Scope | Wide | Circumscribed |
| Duration | 3-5 months | 3-5 days<br>(depending on Kaizen Event) |
| Training | Lean Six Sigma Training<br>(Black Belt; Green Belt or Yellow Belt) | Kaizen specific training, on the job training |
| Team | Multifunctional Team | Multifunctional Team<br>(or sometimes natural teams) |
| Team size | Depending on project | 5-8 people |
| Depth of analysis | Generally high | Generally low |
| Involvement | Part time involvement | Full time involvement |

# The power of *Lean Six Sigma*

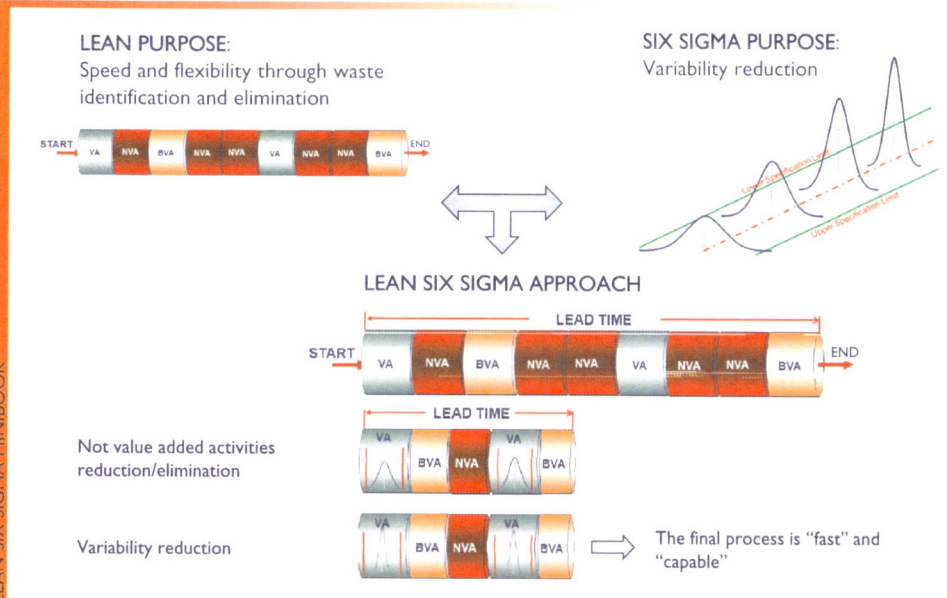

# DEFINE

Arcidiacono G., Calabrese C., Yang K.: Leading processes to lead companies: Lean Six Sigma.
DOI 10-1007/978-88-470-2492-2, © Springer-Verlag Italia 2012

# DEFINE

*Define* phase is the first *step* of a *Lean Six Sigma* project and therefore it is necessary to determine:

- the snapshot of the process through mapping (identifying value added activities and not value added activities)
- the Customer (external or internal)
- weakness and critical points of process
- the scope of the project and therefore the *ring* of intervention
- a measureable indicator from Customer point of view, called *Critical To Quality* (CTQ[G]), and customer satisfaction analysis through a proactive approach instead of reactive
- an estimation of potential benefit (economic and/or strategic, written inside the summary document called *Project Charter* [G]), achievable thanks to the implementation of improvements planned during project development

# SIPOC Diagram

## Objective:

- SIPOC[G] Diagram maps the process from a macroscopic point of view. It should be used during the early stages of an improvement project in order to capture sufficient detail to be able to manage the process

## Overview:

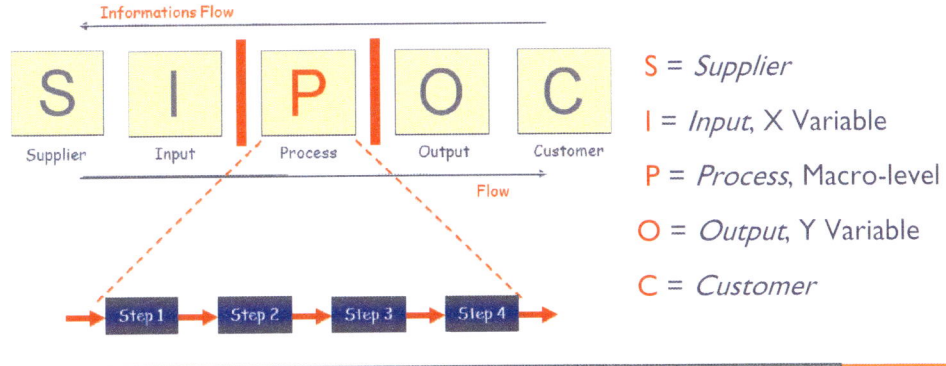

S = *Supplier*

I = *Input*, X Variable

P = *Process*, Macro-level

O = *Output*, Y Variable

C = *Customer*

# SIPOC Diagram

The six steps to construct SIPOC:

1. Identify *Customers* (G) (external and/or internal)
2. Identify process *Outputs*
3. Locate Macro-process boundaries (start and end of process)
4. Determine *Process Owner* (G)
5. Define process *Inputs* and *Relative Suppliers*
6. Repeat the procedure using a "TOP-DOWN" approach (start from the macroprocess to reach more detailed analysis). During the process mapping, distinguish value added phases (VA) from not value added ones (NVA) (see page 19)

LEAN SIX SIGMA MINIBOOK

# SIPOC Diagram

Useful questions during SIPOC construction:

- Who is the *Customer*?

- Is the *Customer* inside or outside the company?

- What does the *Customer* need from the process?

- What are the *Outputs* of the process?

- Where does the examined *Process* start and end (project *ring*)?

- Does the process describe the "as-is" situation or the process desired?

- What are the *Inputs*? What are the specifications agreed upon in terms of *Inputs* with *Suppliers*?

- Who are the *Suppliers*?

# Process Mapping

## Objective:

- The process mapping will describe, at a high level, the analyzed process to identify critical points, both value added and not value added activities

## Overview:

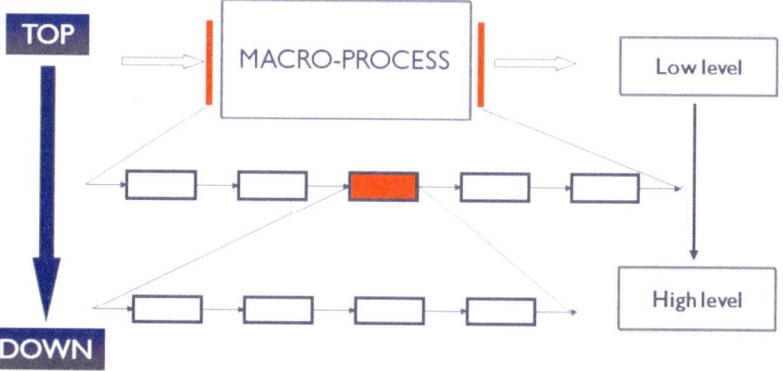

# Process Mapping: Basic flow diagram

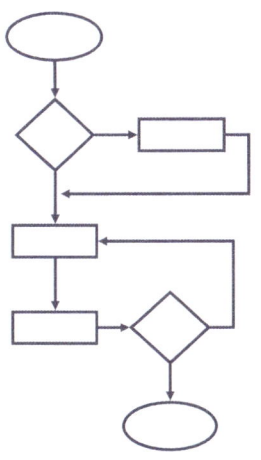

Objective:

- identify the main steps of a process: starting and final points

- identify the decision-making cycles in the process

| ICONS FOR REPRESENTATION | |
|---|---|
| ◇ | Question mark |
| → | Direction of logical flow |
| ▱ | Input/Output |
| ▭ | Process phase |
| ⬭ | Start/Finish of process |

# Process Mapping: Functional flow diagram

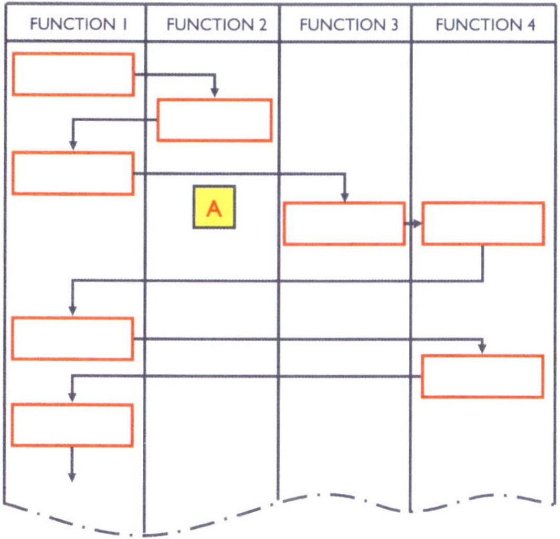

| FUNCTION 1 | FUNCTION 2 | FUNCTION 3 | FUNCTION 4 |

Objective:

- to highlight links among depart-ments/functions in a process flow

- clarify the roles and responsibilities

The horizontal line represents the transfer of functional respon-sibility

# Value Added and Not Value Added

- In a process it is possible to identify three main kinds of activities:

  - ■ Value Added Activity (VA[(G)]):
    - Activity that increases the value of the product/service from the customer's point of view
    - Something that the customer is willing to pay for
  - ■ Not Value Added Activity (NVA[(G)]):
    - Activity that does not add any value to the product/service
  - ■ Business Value Added Activity (BVA[(G)]):
    - Activity that does not add any value to the product/service but is necessary from a business operations' point of view

"To be removed"   "To be enhanced"   "To be reduced"

Example of a company process

# Process Mapping: Activity flow diagram

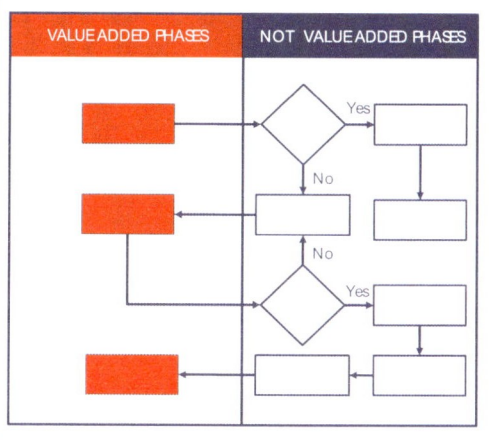

| VALUE ADDED PHASES | NOT VALUE ADDED PHASES |
|---|---|

## Objective:

- to visualize process complexity
- to identify decisional cycles and bottlenecks
- to determine Not Value Added phase for product/service (waste[G])

### Value added phase (VA)

- Customer willing to pay for it
- Step that physically transforms the product/service
- Phase that produces *right first time* product/service

### Not value added phase (NVA)

- Non essential phases for output production:
  - Defects, reworks
  - Testing
  - Checking
  - Transporting, motion, waiting
  - Overproduction, Inventory

# Waste Walk

## Objective:

- The Waste Walk aims to consider the flow of a product / service from start to finish (Example from raw materials to finished goods) and highlights all forms of waste along the way

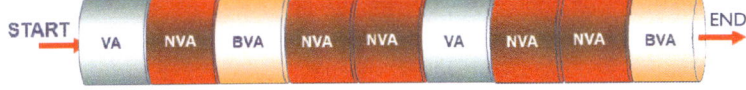

## When to use it:

- The method is very useful to highlight the macro-waste in a process and then to start an improvement action plan with low investment and high return activities on the business processes

# *Waste Walk*: operative procedure

How to perform a "waste walk":

1. Select a product/service provided
2. Go through the end-to-end process following the product and pay attention to detail
3. Identify the waste referring to the eight categories of Lean methodology
4. Collect on a piece of paper all the information related to the waste detected
5. If possible, try to quantify the entity of the waste (€, time, resources, etc.)
6. Clearly describe how wastes are created in the process
7. Generate possible ideas for waste elimination or reduction

# *Waste Walk*: operative procedure

Example of a format for collection of necessary information during a Waste Walk analysis:

| "Waste Walk Format" | | | | |
|---|---|---|---|---|
| Company: _____ Observer: _____ | | Product/Practice followed: _____ Date: _____ | | |
| **Waste category** | **Process Step** | **Waste Comment** | **Waste estimation** | **Possible idea/solution** |
| Insert the waste category identified | Insert the process step where the waste has been identified | Waste description | Try to insert, if it's possible, an estimation of waste entity | Generate one or more ideas for waste reduction/elimination |
| | | | | |
| | | | | |

# *Waste Walk:* operative procedure

## Example of data collection:

| "Waste Walk Format" | | | | |
|---|---|---|---|---|
| Company: XXX | | | Analyzed Product: Az345 | |
| Observer: Mr. Yellow | | | Date: November 20, 2008 | |
| **Waste category** | **Process Step** | **Waste Comment** | **Waste estimation** | **Possible idea/solution** |
| Inventory | Heat treatment | Inventory of treated materials out off the shelves | 14 pallets | Batch size reduction |
| Waiting | Molding | The process is waiting material from machining processes | 10 minutes per hour, 2 operators | |
| Scraps | Assembly | 10% of the parts do not pass the final inspection and need a rework process | 10% of 4000 parts x 6 mins = 2400 mins | |
| Motion | Final assembly | The test station is located 10 meters away from the workplace, workers take 30 seconds to go and come back every time a test is done | 30 secs x 1.100 parts per days = 550 mins | Move the location of the test area |
| Inventory | Final assembly | Too many pallets are waiting for the box closure because of lack of additional elements | 19 pallets | Assemble components only when all objects are present |

# Spaghetti Diagram

## Objective:

- The Spaghetti Diagram is a visual description of actual flows (usually material and/or people flows). It is an effective tool to see the shortcomings in a process such as excessive material movements, long walking distances and layouts that are not designed based on actual process needs

## Overview:

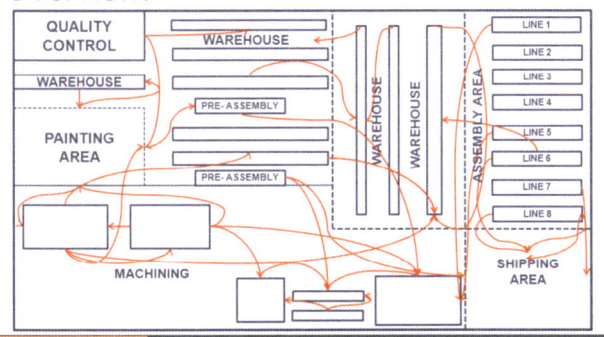

LEAN SIX SIGMA MINIBOOK

# Spaghetti Diagram

How to build a Spaghetti Diagram:

1. Identify the product, file, information, person and material to follow through
2. Take a pen and the facility layout, drawn on paper (or layout of the area in question)
3. Follow the product, information and people flow step by step and draw the actual movement paths on a piece of paper and measure the path distances

Example of a file managed through a service company process

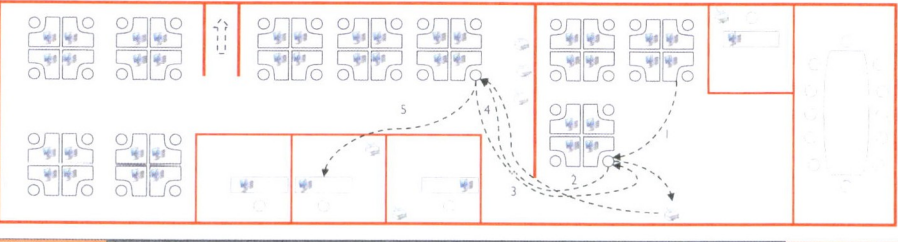

# Spaghetti Diagram

Example of Spaghetti Diagram before and after an optimization activity on Set-Up process (SMED – Single Minute Exchange of Die)

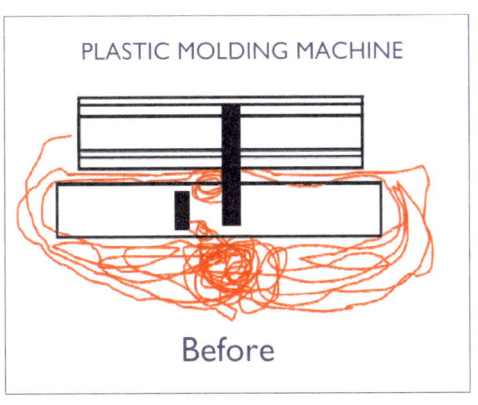

PLASTIC MOLDING MACHINE

Before

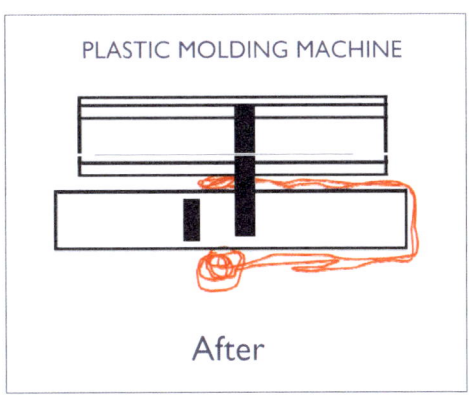

PLASTIC MOLDING MACHINE

After

——— Movement of two operators during a set-up activity

# Product Family Matrix

Objective:

- Product Family Matrix is a tool used to group many products into the "product family". The product will belong to a family if the processes or components necessary to produce it are similar to those of other products in that family

When to use it:

- The tool is used before performing a value stream analysis. When product variety is low, it is possible to have one Value Stream Map for each kind of product. However, in most companies there are hundreds of products and it is impossible to draw hundreds of Value Stream Maps. In this case it is useful to group products into "families"

# Product Family Matrix

## How to build a Product Family Matrix:

1. Place different kinds of product line by line
2. Place processes or components according to product type in columns (pay attention to the downstream processes that generally differentiate the product)
3. For each product, highlight the processes or components used
4. Identify the "Product Family" by degree of similarity in columns used

## Overview:

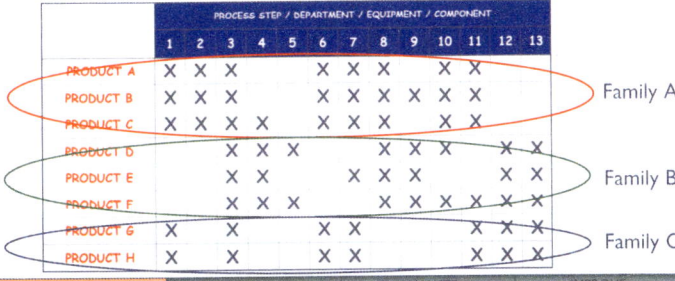

Don't spend a lot of time in Product Family Matrix construction; it is not a value added activity for the customer

It is advisable to build up a Value Stream Mapping for each family, starting from the biggest family in terms of quantity sold to the customer

LEAN SIX SIGMA MINIBOOK

# Value Stream Mapping (VSM)

## Objective:

- Value Stream Mapping (VSM[(G)]) is a diagram of every step involved in the material and information flow necessary to bring the product/service from the order to delivery phase. Value Stream Mapping is a very good starting point to identify opportunities for improvements

## Overview:

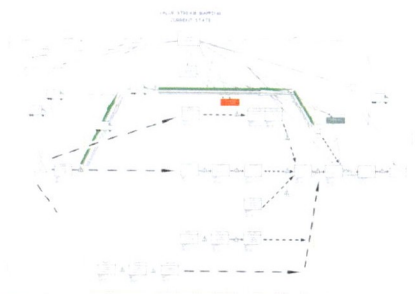
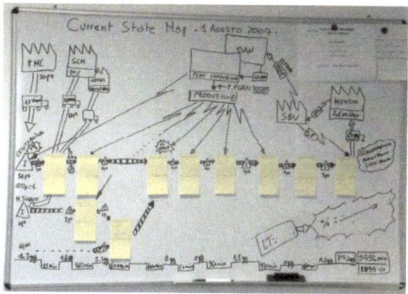

# Value Stream Mapping (VSM): Roadmap

## How to perform a Value Stream analysis:

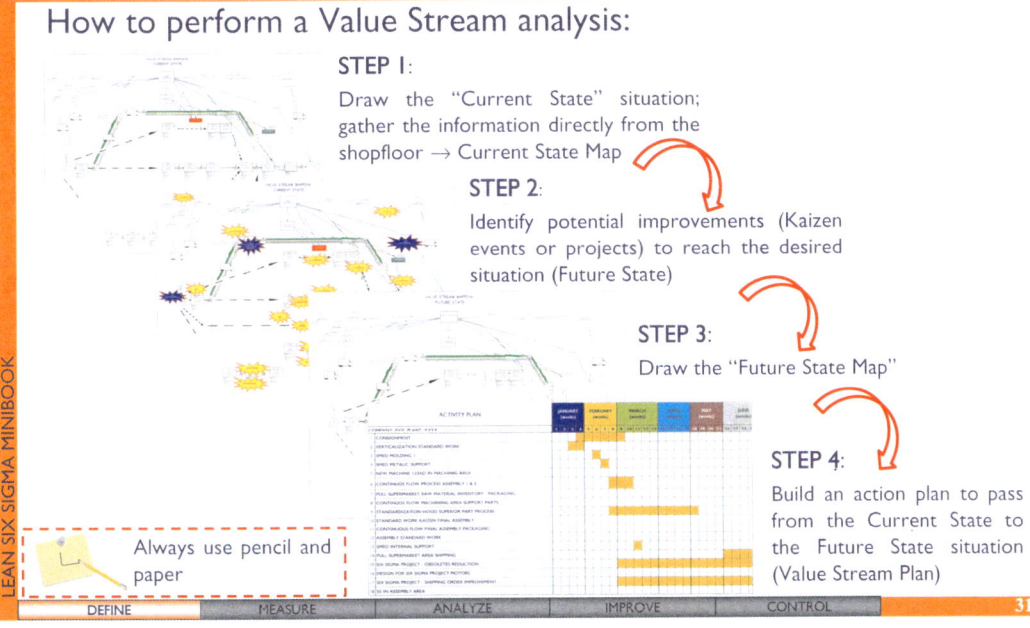

**STEP 1**:
Draw the "Current State" situation; gather the information directly from the shopfloor → Current State Map

**STEP 2**:
Identify potential improvements (Kaizen events or projects) to reach the desired situation (Future State)

**STEP 3**:
Draw the "Future State Map"

**STEP 4**:
Build an action plan to pass from the Current State to the Future State situation (Value Stream Plan)

Always use pencil and paper

# Value Stream Mapping (VSM): Current State

Guidelines to build a Value Stream Map:

1. Identify the product family to analyze (use, if necessary, the Product Family Matrix)

2. Go to the shopfloor and begin mapping, starting from the customer, and go back upstream through the entire flow. Describe step by step processes and gather information such as:

    - Customer information
    - Intermediate inventories and their location
    - How the information flow runs inside the company
    - Production input of the single process
    - Process Cycle Time, Set-Up time, Number of operators for each step, WIP
    - Lead Time[G]
    - Supplier information

# Value Stream Mapping (VSM): Current State

## Example of a Current State Map:

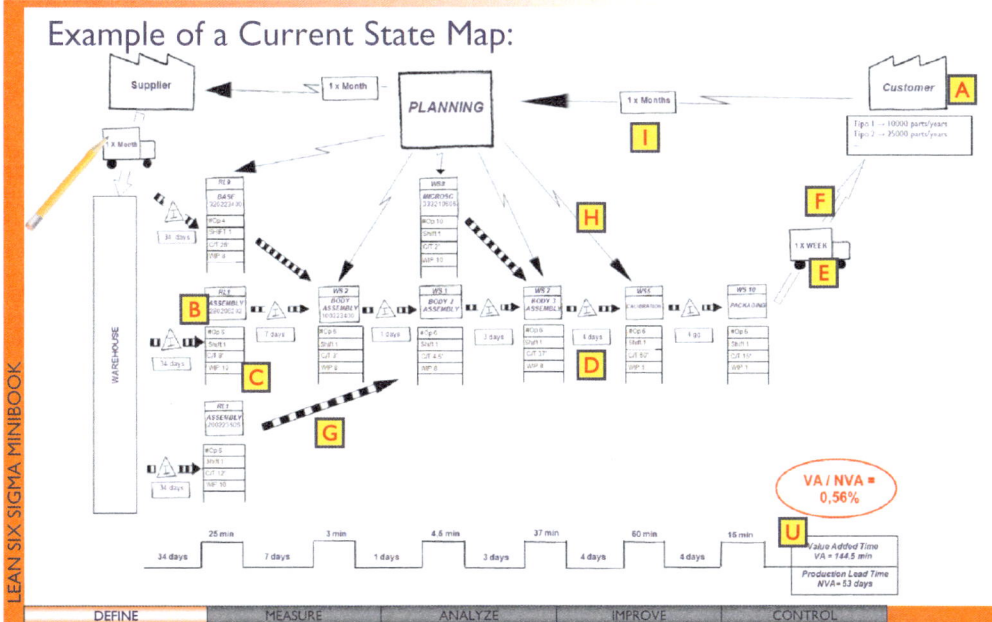

# Value Stream Mapping: Standard Icons

These icons are the common language to do the Value Stream Analysis

| ICON | NAME | DESCRIPTION | PRACTICAL NOTES |
|---|---|---|---|
| XYZ — Prod 1 - 10000 parts/years, Prod 2 - 25000 parts/years | Customers/Suppliers (page 33) **A** | Use to show Customers, Suppliers and External processes. Place in the top right/left corner of the diagram | Gather information such as: •Customer/Supplier Name •Customer/Supplier location (Europe, USA etc.) •Number of products required •Mix of products •Batch size/delivery lead time (supplier) |
| STEP1 NAME OF PROCESS | Process Step **B** | Use to show a process, (exclude elementary work tasks) | Identify Process Name Identify kind of process |
| | Data Box **C** | Use to show all important information concerning each process/customer/supplier. This is generally placed under each process | Example of information for a process: • No. of Operators (VSM icon ◯ ) • Cycle Time/Set Up Time/Uptime • No. of Shifts • Batch Size • Wip • % Defects (PPM,DPMO) |

# Value Stream Mapping: Standard Icons

| ICON | NAME | DESCRIPTION | PRACTICAL NOTES |
|------|------|-------------|-----------------|
| 600 pieces <br> $\boxed{\underline{I}}$ <br> **4.4** days <br> **D** | Inventory | Use to show inventory; try to quantify the inventory in terms of parts, inventory value and time | How to calculate time; example: <br> • average Customer requests → **30,000** parts/year <br> • average Customer demand per day → 30000/220 (220 days in 1 year) = 136 parts/day <br> • inventory time: 600/136= **4.4 days** |
| 2 X Week <br> 1 X Month <br> 1 X Month <br> **E** | Shipment Method | Use to identify kind of shipment | Identify: <br><br> • shipment method (truck; airplane; ship) <br> • delivery frequency <br> • date of shipment |

# Value Stream Mapping: Standard Icons

| ICON | NAME | DESCRIPTION | PRACTICAL NOTES |
|---|---|---|---|
| | External transportation **F** | Movement of finished goods to the customer or movement from supplier to the company | This icon is used if a supplier is involved during the production cycle |
| | Movement of material in push logic **G** | Material is produced and moved to the next step process when not needed | This happens when production is based on schedule but not on Customer demand |
| | Electronic Information **H** | Communication between processes in an electronic flow | Example: <br> • IT system information |
| | Manual information | Use to describe a manual information flow | Example: <br> • production schedule <br> • delivery schedule |
| Weekly Schedule | Information **I** | Use to identify information | Identify the frequency of the information |

# Value Stream Mapping: Standard Icons

| ICON | NAME | DESCRIPTION | PRACTICAL NOTES |
|------|------|-------------|-----------------|
| 3 days | Buffer or safety stock | Used to indicate the buffer/safety stock | The level of buffer stock must be measured (in days or number of parts) under the icon |
| | Supermarket  L | A controlled inventory of parts that "pull" the production from an upstream process | Replenishment is based on actual consumption of stock. If the supermarket is not used, the upstream process doesn't produce any product on supermarket (Pull System) |
| | Withdrawal (Pull material)  M | Used to represent a pull material | Used usually for a request from a process to a supermarket/warehouse |
| -- FIFO --> | First In First Out Sequence | Used to represent transfer of material between processes in a "FIFO" sequence | The first part that goes into the process is the first part to go out. (Example: Conveyor) |

# Value Stream Mapping: Standard Icons

| ICON | NAME | DESCRIPTION | PRACTICAL NOTES |
|------|------|-------------|-----------------|
| | Withdrawal Kanban **N** | Used to instruct the material handler to take and transfer parts from a supermarket/inventory to a downstream process | |
| | Production Kanban **O** | Used to engage the upstream process to produce what is necessary for replenishment of supermarket/inventory | 1 - The upstream process produces what has been consumed from the supermarket<br>2 - The downstream process withdraws from supermarket what it needs in the right quantity |
| | Signal Kanban **P** | This signal is used when the reorder point (G) is reached and a new batch must be produced | The "batch kanban" is necessary when supplying process must produce in batches because of Changeovers |
| | Kanban Post **Q** | Used to show place where Kanban are collected to be taken from material handler to replenish processes | The kanban post could be a mailbox which collects the kanban card or a place which collects bins/balls, etc. |

# Value Stream Mapping: Standard Icons

| ICON | NAME | DESCRIPTION | PRACTICAL NOTES |
|---|---|---|---|
| $\boxed{O X O X}$ | Level load production **R** | The icon is used to indicate levelling of production quantity rather than batching | Tool to intercept batches of kanban and levelling the volume and mix over a period of time ("Every part every day") |
| (glasses icon) | Visual Checking and scheduling **S** | Used to highlight a point where the inventory level is checked visually | Gathering of information through visual checks. Visual management is one of the most important and frequent concepts in lean deployment |
| (starburst icon) **IMPROVEMENT** | Improvement activity **T** | Used to highlight improvement at a specific point of the process stream | To get a global view of what is needed to reach a lean flow. The improvement can be Kaizen Workshop, Six Sigma project, Problem Solving activities, etc. |
| (timeline icon) 25 min VA / 8.5 days NVA / 1 days NVA / Lead Time **U** | Timeline | The timeline shows value added times (Cycle Times) and non-value added times. Use this to calculate Lead Time and Total Cycle Time | The timeline is used to calculate the **Process Cycle Efficiency**, i.e. the ratio between Value-Added Activities and Not Value Added Activities: $PCE = \frac{\sum VA}{\sum NVA} = \frac{25 \text{ min}}{9.5 \text{ days}} = 0.54\%$ |

# Value Stream Mapping (VSM): Future State

What does "Lean" do in a value stream? Key elements for a good future state map design:

1. Identify and eliminate sources of overproduction
2. Produce product/service with *takt-time* rhythm  (identify Takt Time)
3. Create continuous flow, where possible
4. Where a continuous flow is not possible, use pull system for process connection instead of a push system (using supermarket)
5. Reduce batch size and, where possible, introduce the "one piece flow" approach
6. Send the scheduling to the process nearest the Customer (*Pace-Maker* process). Will the customer pull from a finished goods supermarket or directly from shipping?
7. Level mix / volume production at the Pace-Maker process
8. Identify potential improvements for each process step (Change-over, Standard Work, TPM, Variability reduction, Kanban, Cell Design, Problem Solving)

# Value Stream Mapping (VSM): Future State

## How to reach the Future State Map:

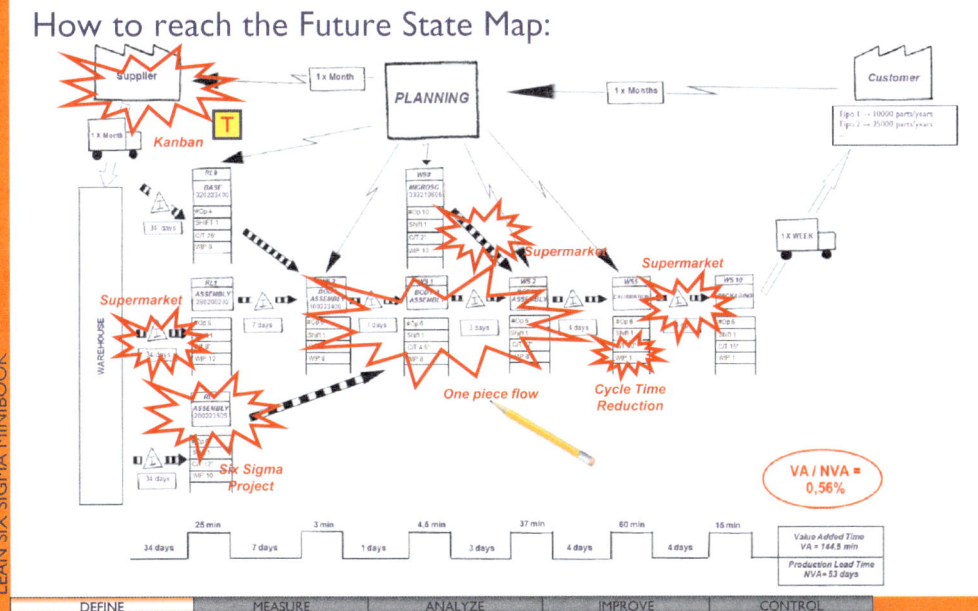

LEAN SIX SIGMA MINIBOOK

# Value Stream Mapping (VSM): Future State

Example of a Future State Map:

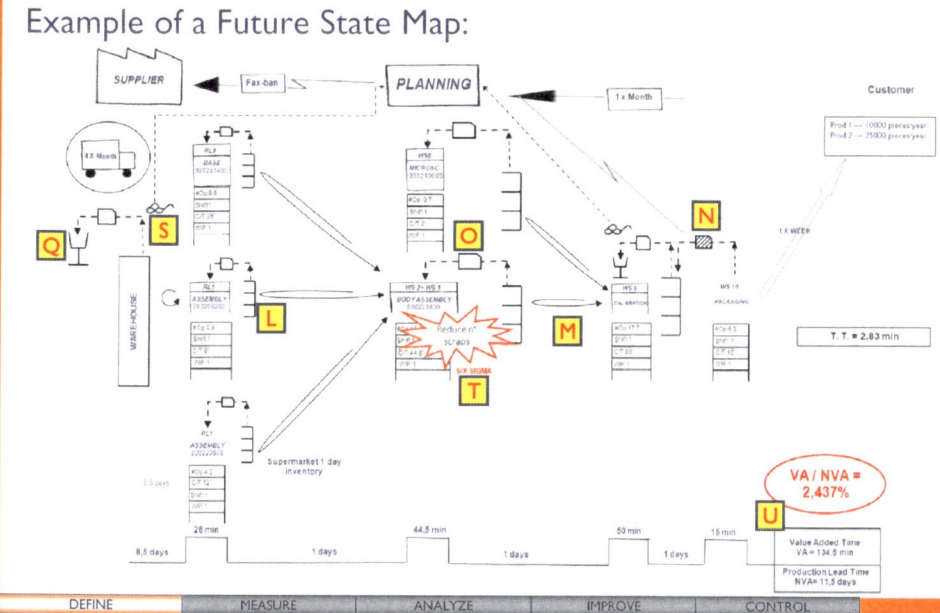

# Value Stream Mapping (VSM): Future State

Example of a Value Stream Map:

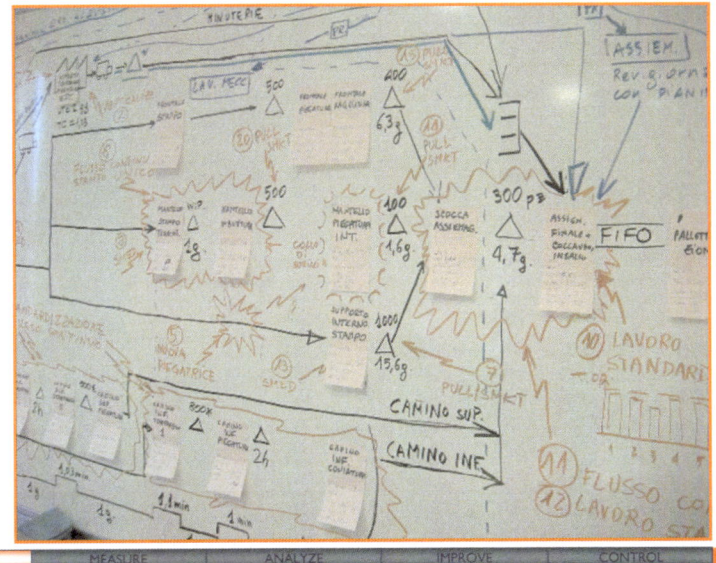

# CTQ-Tree Diagram

## Objective:

- Tree Diagram is a tool moving from VOC[G] (*Voice Of the Customer*) to one or more CTQs, translating the customer's voice into objectively measurable indicators.

## Overview:

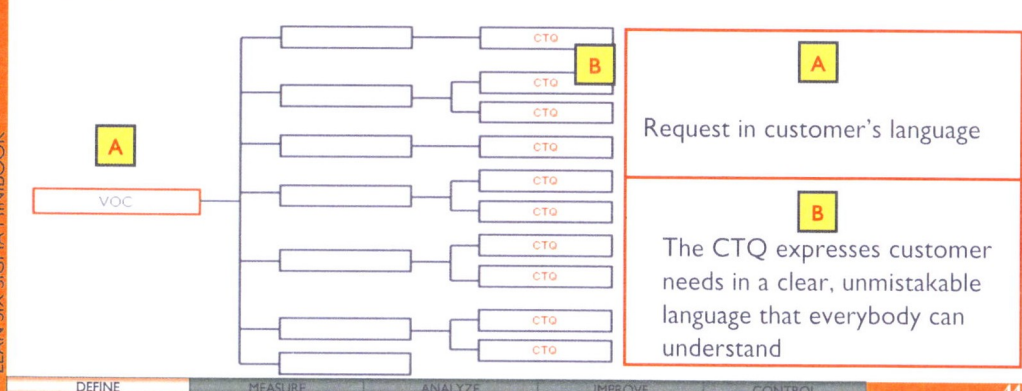

# Kano Diagram

## Objective:

- Kano Diagram is a tool that aims to identify the important aspects to drive *Customer Satisfaction* (G)

## Overview:

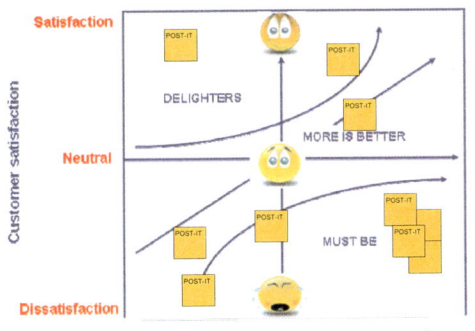

**Must Be:** Qualities that are present and what customers expect to be present (e.g. product or service must be safe)

**More is Better:** These qualities are directly correlated to customer satisfaction. The more present they are, the bigger the Customer Satisfaction

**Delighters:** Qualities unexpectedly found by the Customer and consequently, if present, increase his/her satisfaction

# Project Charter

| Project Charter | | | |
|---|---|---|---|
| Title | Reduce the percentage of scrap in the cable assembly lines | | |
| Scope | The project will focus on defect E, G, A, which constitute 75.3% of the total defects examined | | |
| Team Leader | M. White (GB) | Telephone | 3481 |
| Team Member | P. Green | Telephone | 3456 |
| | S. Yellow | Telephone | 3476 |
| | R. Black | Telephone | 3454 |
| | C. Blue | Telephone | 3410 |
| | | Telephone | |
| Process Owner | F. Violet Production Responsible | Telephone | 3400 |
| Champion | F. Orange | Telephone | 3289 |
| Duration | 4 | months | |
| CTQ | N° Scraps/Pieces produced | | |
| Actual value | 12% | | |
| Expected value | 2% | | |
| Savings | 78.000 € | Euro/year | |
| Constraints | Do not increase the cycle time | | |
| Milestone | Expected Start | Expected closure | % Progress |
| Define | 08 february | 22 march | 100% |
| Measure | 22 march | 30 april | 100% |
| Analyze | 30 april | 27 may | 100% |
| Improve | 27 may | 15 june | 100% |
| Control | 15 june | 04 july | 75% |

# Project Charter

## Contents of Project Charter:

| | |
|---|---|
| Title | Indicate project title |
| Scope | Describe in more detail the purpose of the project, in order to better explain the meaning of the title |
| Team | Insert the people who are involved in the project deployment:<br>    – *Team Leader*<br>    – *Team Members*<br>    – *Process Owner:* Owner of the area (Department/Office) involved in the project<br>    – *Champion/Sponsor:* is the financial backer of the project Six Sigma |
| Duration | Identify project duration (4-5 months) |
| CTQ | Identify the project CTQ to be analyzed (it could be one or more than one),  the current value and the expected value (*target*) at the end of the project |
| *Savings* (G) | Insert the value of financial benefits (in terms of revenue and/or cost) that should be achieved when the CTQ target is reached at the end of the project (benefits per year) |
| Constraints | Identify the constraints which must be respected during the project development |
| Phases | Highlight the start-up date for each phase, expected closure date and the real percentage of progress (Project Charter is an "ongoing document" and the starting point of all project meetings) |

# COPQ: Cost Of Poor Quality

- Cost of Poor Quality, called COPQ[G] (*Cost Of Poor Quality*), are those costs due to poor performance of manufacturing and/or transactional processes and include labor costs, energy, materials, depreciation, which must be sustained to avoid generating non-conformity or in response to their occurrence

- A possible model for COPQ could be the following:

| | | |
|---|---|---|
| **COPQ** | COSTS OF QUALITY | • Prevention costs<br>• Checking costs |
| | COSTS OF NON CONFORMITY | • Costs due to internal defects<br>• Costs due to external defects |

# COPQ: Cost Of Poor Quality (example)

| COPQ | COST OF QUALITY | PREVENTION COSTS | - Preventive maintenance<br>- Process Audit<br>- Product Audit<br>- Process Capability study<br>- Machine Capability study |
| | | CHECKING COSTS | - Checking<br>- Test<br>- Check on invoices<br>- Check on purchase order<br>- Life cycle test on products |
| | COST OF NON CONFORMITY | COSTS INTERNAL DEFECTS | - Scraps<br>- Reworks<br>- Delays and waiting<br>- Invoices issued late<br>- Wrong sales forecasts<br>- Production stoppages for problems |
| | | COSTS EXTERNAL DEFECTS | - Customer returns<br>- Warranty assistance<br>- Penalties<br>- Downgrade products<br>- Production of documents related to returns |

# MEASURE

Arcidiacono G., Calabrese C., Yang K.: Leading processes to lead companies: Lean Six Sigma.
DOI 10-1007/978-88-470-2492-2, © Springer-Verlag Italia 2012

# MEASURE

*Measure* phase is the second step of a Lean Six Sigma project, where:

- a "rational" data collection is performed for the scope chosen: this collection requires effective and efficient planning in order to create a database of knowledge to record the process which will highlight the critical issues from an objective standpoint

- the data is interpreted through statistical tools (in case of samples, to test their significance and how they are seen/shown over-viewed)

- the reliability of data is verified

- process performance is calculated through the proper KPI (OEE, Takt Time, Process Sigma, Process Capability etc.)

# Sampling

## Objective:

- Gather a subset of data (n) representative of the population (N)

## When to use it:

- When the observation of all data (population) would:
    - require too many financial resources
    - take longer than the time at one's disposition
    - "destroy" the entire population (e.g. in case of destructive tests)

# Sampling

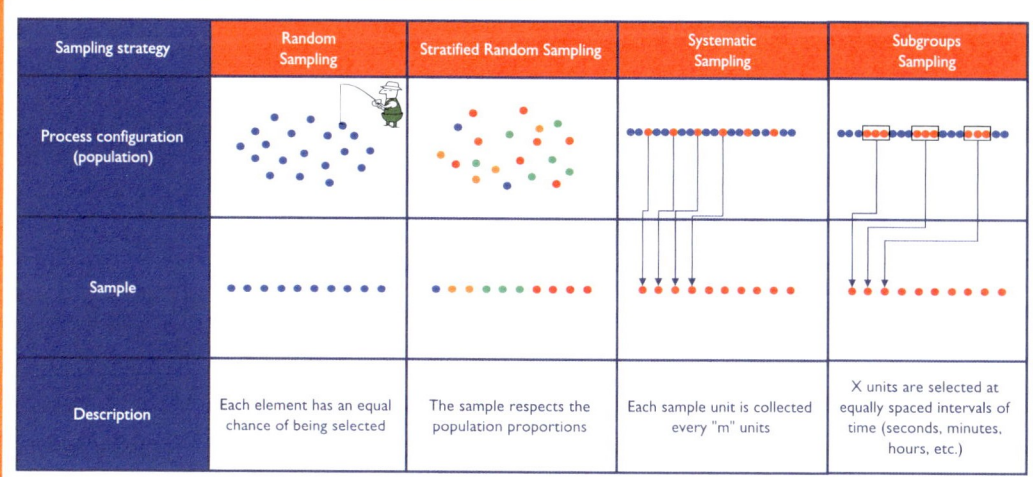

| Sampling strategy | Random Sampling | Stratified Random Sampling | Systematic Sampling | Subgroups Sampling |
|---|---|---|---|---|
| Process configuration (population) | | | | |
| Sample | | | | |
| Description | Each element has an equal chance of being selected | The sample respects the population proportions | Each sample unit is collected every "m" units | X units are selected at equally spaced intervals of time (seconds, minutes, hours, etc.) |

# Sampling

Sample size calculation to estimate the population's mean (formula for continuous data):

$$n = \left(\frac{2s}{d}\right)^2$$  Formula (i)

where:

- $n$ = Sample size
- $s$ = Standard deviation or its estimate
- $2$ = factor corresponding to a Confidence Interval of 95% (the exact value is 1.96)
- $d$ = Precision required in mean estimation

# Sampling

Sample size calculation for the estimation of the population proportion (formula for discrete data):

$$n = \left(\frac{2}{d}\right)^2 (p)(1-p)$$ Formula (ii)

where:

- $n$ = Sample size
- $p$ = Proportion estimation (if it is not known use $p = 0.50$)
- $2$ = factor corresponding to a Confidence Interval of 95% (the exact value is 1.96)
- $d$ = Precision required in proportion estimation

# Sampling

- Formulas (i) and (ii) are valid if the sample size is less than 5% in comparison with population size

$$\frac{n}{N} < 0.05$$

- If the sample size is more than 5% of the size of the population, it is possible to adjust the sample size obtained with (i) and (ii) using the formula below:

$$n_{correct} = \frac{n}{1 + \frac{n}{N}}$$

CORRECTIVE FORMULA

# Basic Statistics

Objective:

- Represent the main statistical properties of a set of data (sample or population)

Characteristics:

- Location parameters
  - Mean, Mode, Median, Quartiles, Percentiles
- Dispersion parameters
  - Range, Standard Deviation, Variance
- Shape parameters
  - *Skewness, Kurtosis*

# Basic Statistics

## Location parameters

The location parameters aim to identify the most frequent values of distributing data

## Dispersion parameters

The dispersion parameters can assess the variability of data

## Shape parameters

The shape parameters are used to assess whether the data collected are arranged according to a symmetrical distribution

# Basic Statistics

| Location parameters: | |
|---|---|
| Mean | $$\overline{X} = \frac{\sum_i X_i}{N}$$ |
| Mode | The value, among data, with highest frequency |
| Median | The value having the characteristic that one half of the values is larger and the other half smaller. The value divides the set of data into two equal parts (it is the mean of the two central values if the number of data is even) |
| First Quartile (Q1) | The value of a concept for which three quarters (75%) of the value are larger and one quarter (25%) is smaller |
| Third Quartile (Q3) | The value of a concept, thus one quarter (25%) of the value is greater and three quarters (75%) are smaller |
| Dispersion parameters: | |
| Range | $$Range = x_{max} - x_{min}$$ |
| Standard Deviation (StDev) | $$\sigma = \sqrt{\frac{\sum (X_i - \overline{X})^2}{N-1}}$$ |
| Variance | $$\sigma^2 = \frac{\sum (X_i - \overline{X})^2}{N-1}$$ |

# Basic Statistics

| Shape parameters: | |
|---|---|
| **Skewness** <br><br> $$Skewness = \frac{N}{(N-1)(N-2)} \sum \left[ \frac{(x_i - \bar{x})}{s} \right]^3$$ | The value that measures the asymmetry of distribution: <br><br> • Skewness < 0: the distribution is shifted to the right <br><br> • Skewness > 0: the distribution is shifted to the left <br><br> • Skewness = 0: the distribution is symmetrical |
| **Kurtosis** <br><br> $$Kurtosis = \frac{N(N-1)}{(N-1)(N-2)(N-3)} \sum \left[ \frac{(x_i - \bar{x})}{s} \right]^4 - $$ $$- \frac{3(N-1)^2}{(N-2)(N-3)}$$ | The Kurtosis is a measure of how the distribution of the analyzed data differs from a normal distribution: <br><br> • Kurtosis < 0: the distribution has a softer peak, shoulders are bigger and tails are thinner than a normal distribution <br><br> • Kurtosis > 0: the distribution has a sharper peak, shoulders are finer, and tails are larger than a normal distribution |

# Basic Statistics

A graphic explanation of the meaning of Skewness, a symmetry parameter, is as follows:

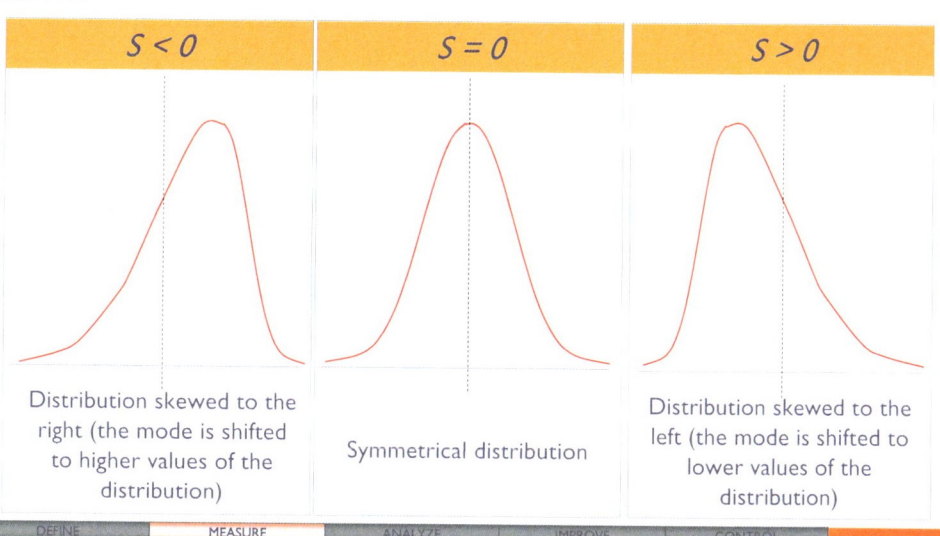

| S < 0 | S = 0 | S > 0 |
| --- | --- | --- |
| Distribution skewed to the right (the mode is shifted to higher values of the distribution) | Symmetrical distribution | Distribution skewed to the left (the mode is shifted to lower values of the distribution) |

# Basic Statistics

A graphic explanation of the meaning of the Kurtosis, a symmetry parameter, is as follows:

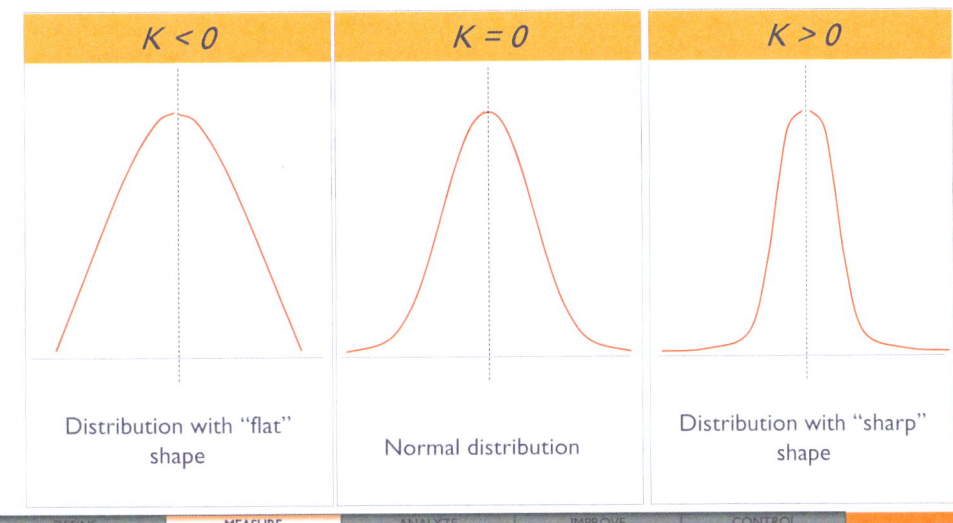

| K < 0 | K = 0 | K > 0 |
|---|---|---|
| Distribution with "flat" shape | Normal distribution | Distribution with "sharp" shape |

# Basic Statistics

MINITAB:

Stat > Basic Statistics > Display Descriptive Statistics...

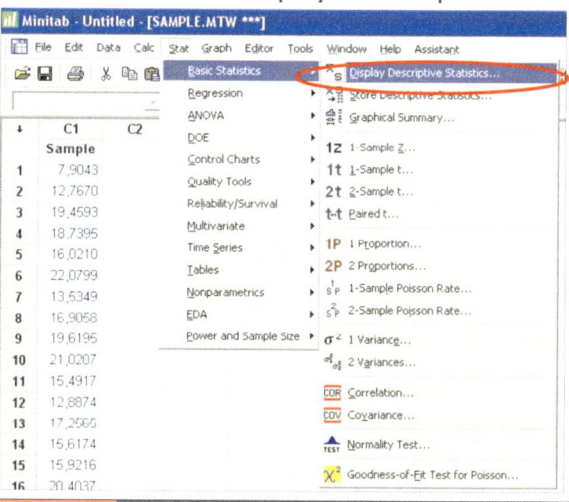

# Basic Statistics

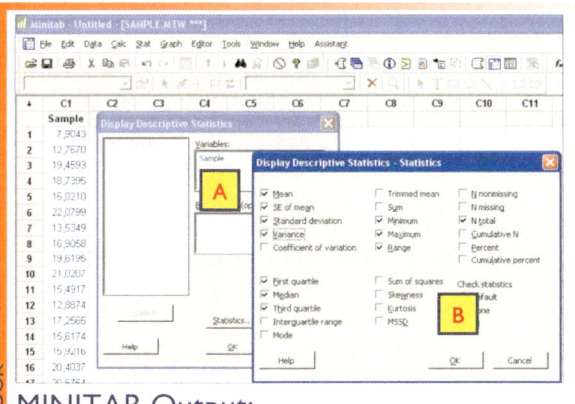

**A**

Insert the column of the sample to be analyzed in terms of basic statistics

**B**

Select all parameters of interest

## MINITAB Output:

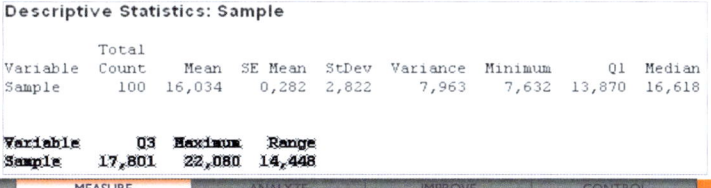

**Descriptive Statistics: Sample**

|  | Total |  |  |  |  |  |  |  |
|---|---|---|---|---|---|---|---|---|
| Variable | Count | Mean | SE Mean | StDev | Variance | Minimum | Q1 | Median |
| Sample | 100 | 16,034 | 0,282 | 2,822 | 7,963 | 7,632 | 13,870 | 16,618 |

| Variable | Q3 | Maximum | Range |
|---|---|---|---|
| Sample | 17,801 | 22,080 | 14,448 |

# Confidence Interval

## What is it?

- The Confidence Interval (CI)[G] is the interval that contains the mean (or proportion, median, standard deviation) of the population with a probability of 95%

## When to use it:

- The Confidence Interval is used to identify whether the analyzed sample belongs to a certain population

## Overview:

Sample mean $\overline{X}$

95% Confidence Intervals

Interval within which is located, with a probability of 95%, the mean of the population from which the sample comes

# Confidence Interval

The formula for determining the width of the Confidence Interval for the mean of a sample is:

$$CI_\mu = \overline{X} \pm 2\frac{s}{\sqrt{n}}$$

where:

$\overline{X}$ = Sample mean

CI = 95% Confidence interval

n = Sample size

s = Sample standard deviation*

Example: Estimation of the mean width of a particular mechanical part

$\overline{X}$ = 21.0 mm
s = 0.25 mm      CI = (21.0 ± 0.098)
n = 25

*when the population standard deviation is known, use it instead of the sample standard deviation (s)

# Confidence Interval

The following formula is used to determine the width of the Confidence Interval for the proportion of a sample:

$$CI_P = \bar{p} \pm 2\sqrt{\frac{\bar{p} \times (1-\bar{p})}{n}}$$

where:

$\bar{p}$ = Sample proportion estimate

CI = 95% Confidence Interval

n = Sample size

Example: Estimate in order to assess the proportion of defects in a specific process

$\bar{p} = 0.15$

$n = 25$

$CI = (0.15 \pm 0.0714)$

# Graphical Summary

Objective:

- The tool aims to give graphical and statistical representation of the parameters found in the collected data

Characteristics (see page 72):

- Histogram with reference curve   ⟶ A
- Normality Test   ⟶ B
- *Basic Statistics*   ⟶ C
- Verification of *Outliers* (G) presence   ⟶ D
- Confidence Interval   ⟶ E

# Graphical Summary

MINITAB:

Stat > Basic Statistics > Graphical Summary...

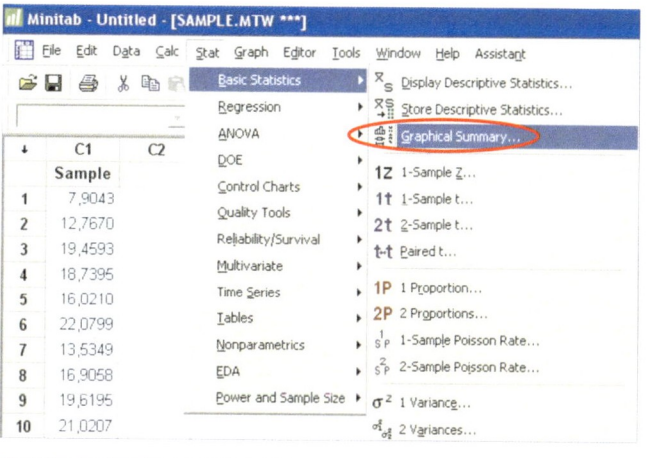

# Graphical Summary

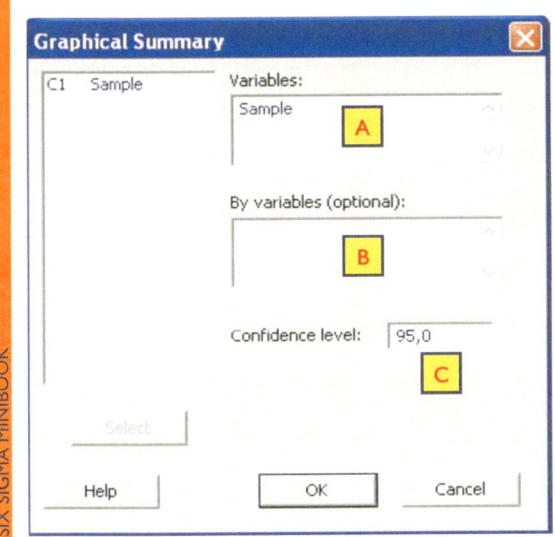

Insert the column containing the sample to be analyzed

Insert potential stratification factors

Choose the level of confidence to determine the width of the Confidence Intervals

LEAN SIX SIGMA MINIBOOK

# Graphical Summary

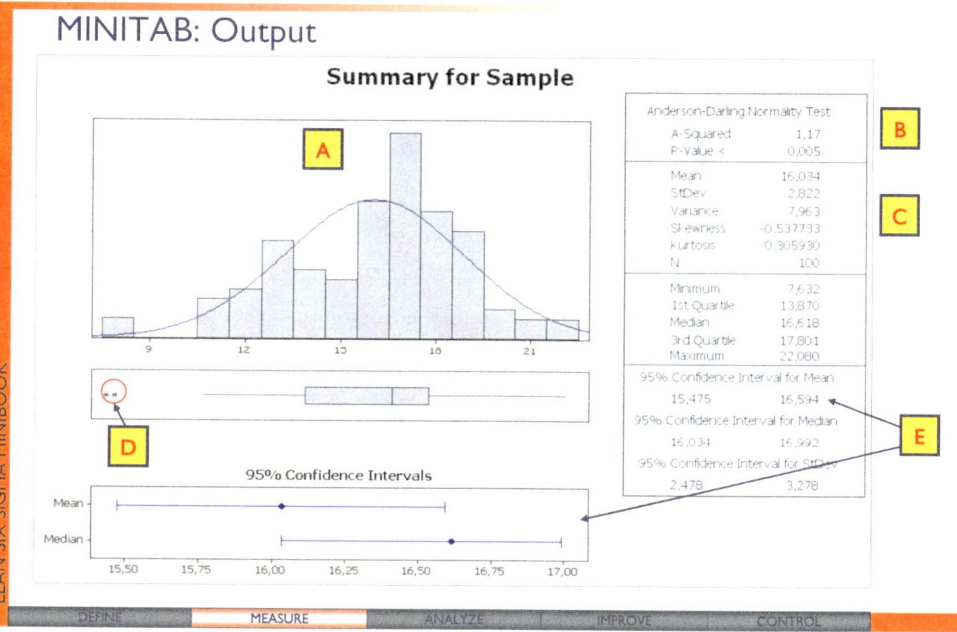

# Graphical Summary

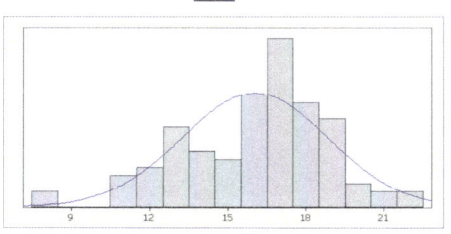

- This is a schematic Bar Chart (Histogram)
- The chart is useful to check on a qualitative way the pattern of data collected in terms of mean, variability and presence of any abnormalities such as outliers, mixture of two distributions, etc.
- This chart shows the normal curve that best fits the data shown

| Anderson-Darling Normality Test | |
|---|---|
| A-Squared | 1,17 |
| P-Value < | 0,005 |

- It reports the results of the Anderson-Darling statistical test in order to verify the normality of the sample:
  - if the *P-Value* [G] is bigger than the threshold value chosen (generally 0.05, that is 5%), it is possible to say that the sample comes from a normal distribution
  - if the P-Value is less than the threshold value chosen, the sample data doesn't respect a normal distribution

# Graphical Summary

| | |
|---|---|
| Mean | 16,034 |
| StDev | 2,822 |
| Variance | 7,963 |
| Skewness | -0,537733 |
| Kurtosis | 0,305930 |
| N | 100 |
| Minimum | 7,632 |
| 1st Quartile | 13,870 |
| Median | 16,618 |
| 3rd Quartile | 17,801 |
| Maximum | 22,080 |

- This section has a summary of data through the usage of the parameters of location, dispersion, symmetry and sample size (see pages 60 and 61)

- **Location parameters**: Mean, Median, First Quartile, Third Quartile

- **Dispersion parameters**: Minimum, Maximum, Standard Deviation (StDev)

- **Shape parameters**: Skewness, Kurtosis

- D is a graph of the sample data through the use of the Boxplot
- The symbols * highlight points that may not belong to the same distribution of the remaining data
- These points are called Outliers and can often be associated with special events

# Graphical Summary

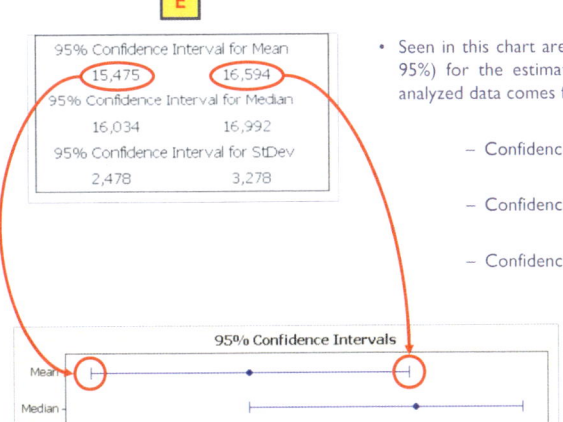

E

95% Confidence Interval for Mean
15,475          16,594
95% Confidence Interval for Median
16,034          16,992
95% Confidence Interval for StDev
2,478          3,278

95% Confidence Intervals

Mean

Median

15,50    15,75    16,00    16,25    16,50    16,75    17,00

- Seen in this chart are confidence intervals (with confidence level of 95%) for the estimation on population, which the sample of the analyzed data comes from:

    – Confidence Interval for mean

    – Confidence Interval for median

    – Confidence Interval for Standard Deviation

# Boxplot

Objective:

- The Boxplot is a tool useful for studying the distribution of collected data and to obtain information on position, dispersion, and symmetry

When to use it:

- With the Boxplot you can see if any Outliers exist (if yes, an investigation must be done to find out the reasons), or if there are any points not belonging to the same distribution of the remaining data, they must be probed, too

# Boxplot

MINITAB:

Graph > Boxplot…

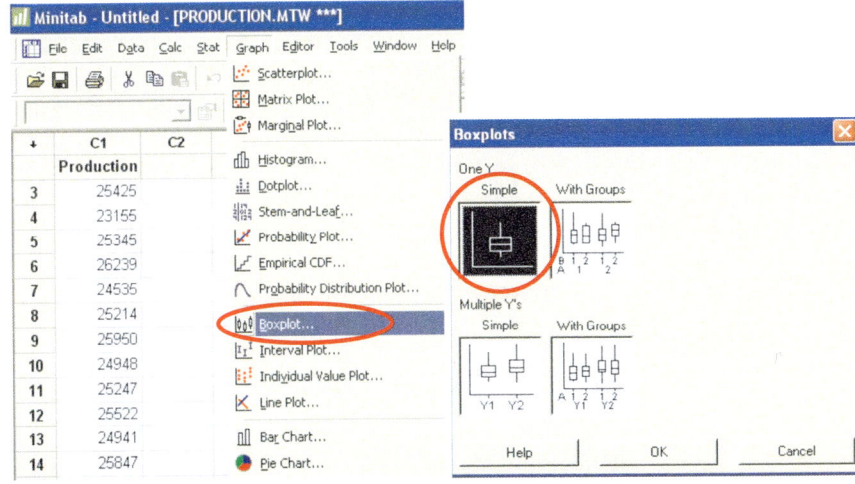

LEAN SIX SIGMA MINIBOOK

# Boxplot

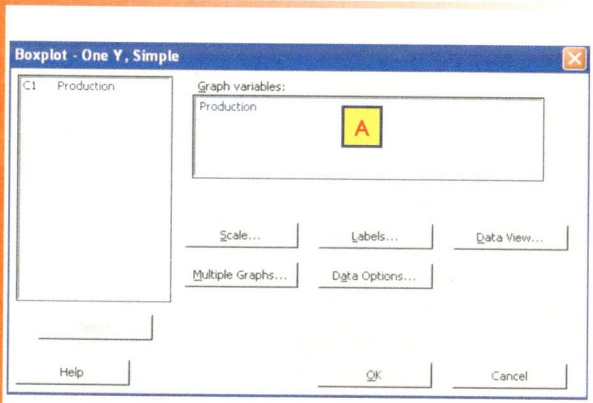

Select the variable stored in a single column to be plotted

# Boxplot

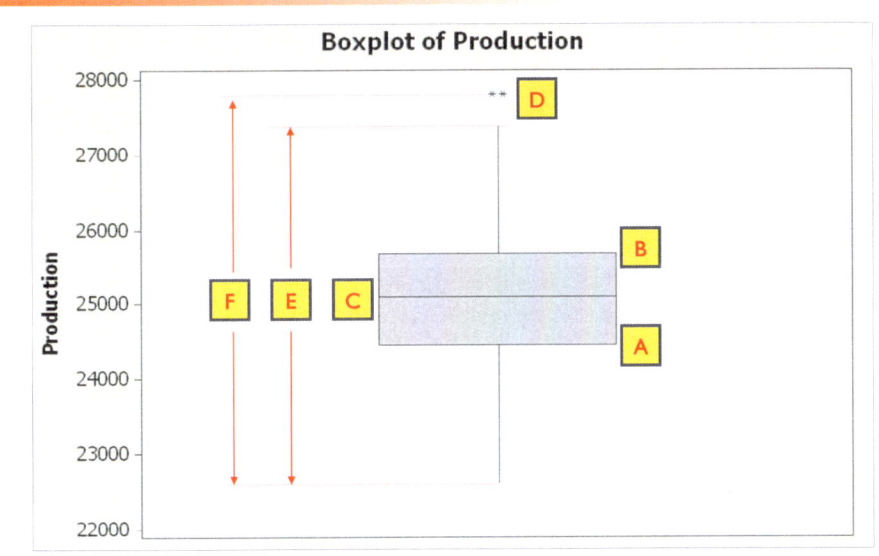

# Boxplot

**A** Lower Quartile or First Quartile ($Q_1$): Cut-off point for lowest 25% of data

**B** Upper or Third Quartile ($Q_3$): Cut-off point for lowest 75% of data (or highest 25% of data)

**C** Median ($Q_2$): Cut-off point for 50% of data (50th percentile)

**D** Outliers: Observations that are numerically distant from the rest of data, i.e. unusually large or small data points (verify why they are present!). Minitab uses the quartile method

$$\begin{cases} LS = Q_3 + 1,5\left(Q_3 - Q_1\right) \\ LI = Q_1 - 1,5\left(Q_3 - Q_1\right) \end{cases} \quad \text{QUARTILE METHOD}$$

**E** Upper whisker and Lower whisker: These are estimates of the upper and lower limits of the data set (excluding outliers)

**F** Range: Maximum – Minimum

# Gage R&R

Objective:
- Gage R&R is a measurement of the capability of a measurement system to obtain the same measurement reading consistently with repeated measurement takings. Gage R&R decomposes the total variation in measured data into part to part variation and measurement variation, and determines the capability of the measurement system by comparing measurement variation vs total variation

Components of Variation :

$$\sigma^2_{total} = \sigma^2_{part-to-part} + \sigma^2_{measurement}$$

$$\sigma^2_{repeatability} + \sigma^2_{reproducibility}$$

The measurement system is valid if the greater part of variability is attributable to the process ($\sigma_{part\ to\ part}$)

LEAN SIX SIGMA MINIBOOK

# Gage R&R

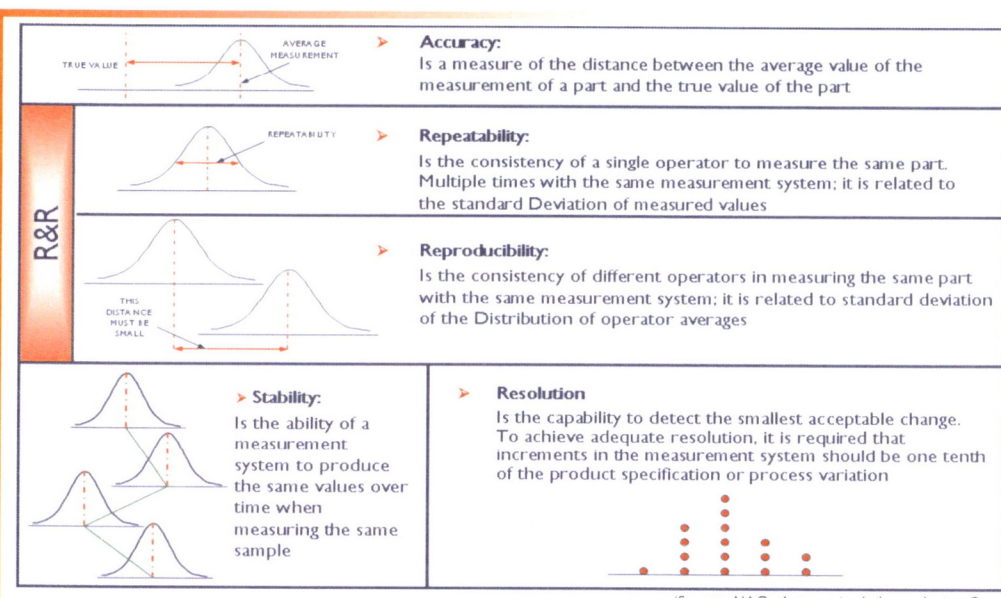

**R&R**

> **Accuracy:**
> Is a measure of the distance between the average value of the measurement of a part and the true value of the part

TRUE VALUE — AVERAGE MEASUREMENT

> **Repeatability:**
> Is the consistency of a single operator to measure the same part. Multiple times with the same measurement system; it is related to the standard Deviation of measured values

REPEATABILITY

> **Reproducibility:**
> Is the consistency of different operators in measuring the same part with the same measurement system; it is related to standard deviation of the Distribution of operator averages

THIS DISTANCE MUST BE SMALL

> **Stability:**
> Is the ability of a measurement system to produce the same values over time when measuring the same sample

> **Resolution**
> Is the capability to detect the smallest acceptable change. To achieve adequate resolution, it is required that increments in the measurement system should be one tenth of the product specification or process variation

(Source: AIAG, *Automotive Industry Action Group*)

# Gage R&R (Continuous Data)

## Gage R&R Test Plan (Manual):

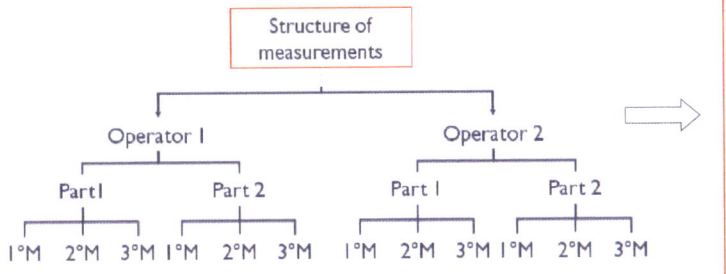

| Parts | Operators | Measurements |
|---|---|---|
| 1 | Operator 1 | 97,500 |
| 1 | Operator 1 | 97,870 |
| 1 | Operator 1 | 97,516 |
| 1 | Operator 1 | 97,685 |
| 1 | Operator 2 | 98,055 |
| 1 | Operator 2 | 97,870 |
| 1 | Operator 2 | 98,018 |
| 1 | Operator 2 | 97,685 |
| 2 | Operator 1 | 100,140 |
| 2 | Operator 1 | 100,088 |
| 2 | Operator 1 | 100,236 |
| 2 | Operator 1 | 99,940 |

1. Each operator will measure each part multiple times (recommended 3 times)
2. The data must be balanced, i.e., each operator must measure the same part equal number of times
3. The selected parts must be representative of the range of variation of the process
4. Operators should carry out the tests "blindly", i.e. without being influenced by other operators' measurements, and sequence their measurement jobs randomly

# Gage R&R (Continuous Data)

Gage R&R Test Plan (Automatically Created by Minitab):

MINITAB:

Quality Tools > Gage Study > Create Gage R&R Study Worksheet...

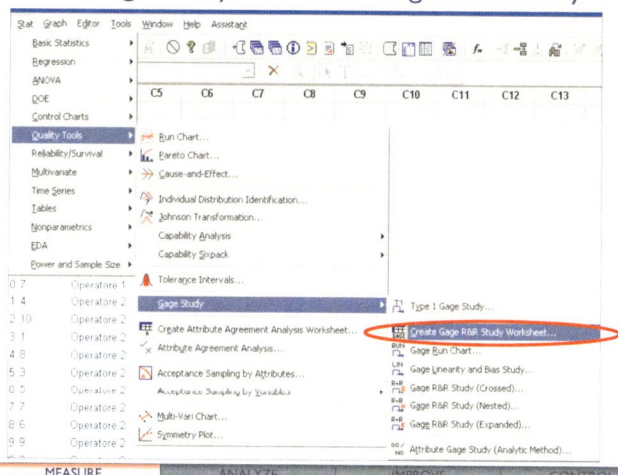

# Gage R&R (Continuous Data)

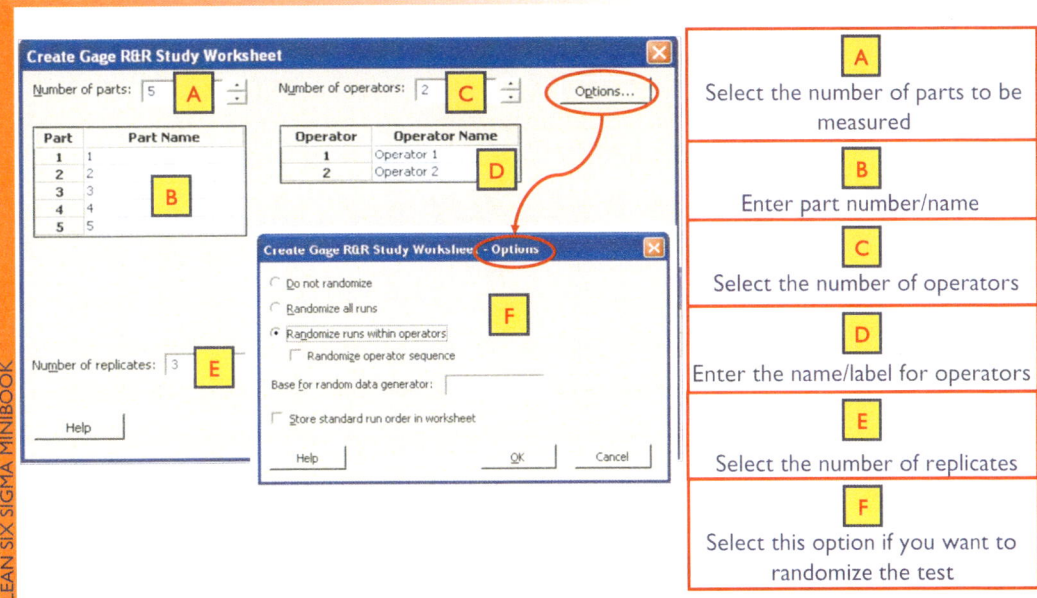

LEAN SIX SIGMA MINIBOOK

**Create Gage R&R Study Worksheet**

Number of parts: 5    A

Number of operators: 2    C    Options...

| Part | Part Name |
|------|-----------|
| 1 | 1 |
| 2 | 2 |
| 3 | 3 |
| 4 | 4 |
| 5 | 5 |

B

| Operator | Operator Name |
|----------|---------------|
| 1 | Operator 1 |
| 2 | Operator 2 |

D

Number of replicates: 3    E

Help

**Create Gage R&R Study Worksheet - Options**

○ Do not randomize
○ Randomize all runs
● Randomize runs within operators
  ☐ Randomize operator sequence

Base for random data generator:

☐ Store standard run order in worksheet

Help        OK        Cancel

F

| | |
|---|---|
| **A** | Select the number of parts to be measured |
| **B** | Enter part number/name |
| **C** | Select the number of operators |
| **D** | Enter the name/label for operators |
| **E** | Select the number of replicates |
| **F** | Select this option if you want to randomize the test |

# Gage R&R (Continuous Data)

## MINITAB: Output

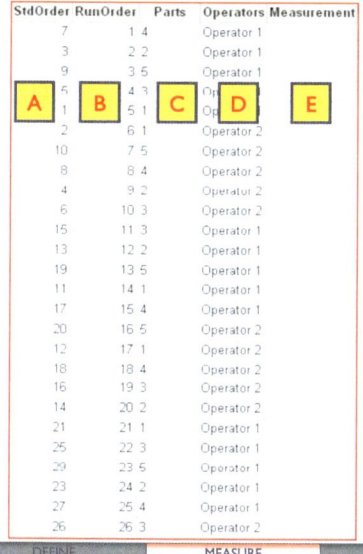

| StdOrder | RunOrder | Parts | Operators Measurement |
|---|---|---|---|
| 7 | 1 | 4 | Operator 1 |
| 3 | 2 | 2 | Operator 1 |
| 9 | 3 | 5 | Operator 1 |
| 5 | 4 | 3 | Operator 1 |
| 1 | 5 | 1 | Operator 1 |
| 2 | 6 | 1 | Operator 2 |
| 10 | 7 | 5 | Operator 2 |
| 8 | 8 | 4 | Operator 2 |
| 4 | 9 | 2 | Operator 2 |
| 6 | 10 | 3 | Operator 2 |
| 15 | 11 | 3 | Operator 1 |
| 13 | 12 | 2 | Operator 1 |
| 19 | 13 | 5 | Operator 1 |
| 11 | 14 | 1 | Operator 1 |
| 17 | 15 | 4 | Operator 1 |
| 20 | 16 | 5 | Operator 2 |
| 12 | 17 | 1 | Operator 2 |
| 18 | 18 | 4 | Operator 2 |
| 16 | 19 | 3 | Operator 2 |
| 14 | 20 | 2 | Operator 2 |
| 21 | 21 | 1 | Operator 1 |
| 25 | 22 | 3 | Operator 1 |
| 29 | 23 | 5 | Operator 1 |
| 23 | 24 | 2 | Operator 1 |
| 27 | 25 | 4 | Operator 1 |
| 26 | 26 | 3 | Operator 2 |

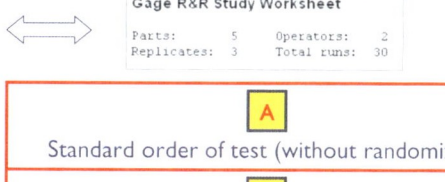

Gage R&R Study Worksheet

| Parts: | 5 | Operators: | 2 |
|---|---|---|---|
| Replicates: | 3 | Total runs: | 30 |

### A
Standard order of test (without randomization)

### B
Randomized test run order

### C
Part number to be measured in each run

### D
Operator number/name

### E
Measurement data to be entered

LEAN SIX SIGMA MINIBOOK

DEFINE — MEASURE — ANALYZE — IMPROVE — CONTROL

86

# Gage R&R (Continuous Data)

Analysis of Gage R&R with MINITAB:

Stat > Quality Tools > Gage Study > Gage R&R Study (Crossed)…

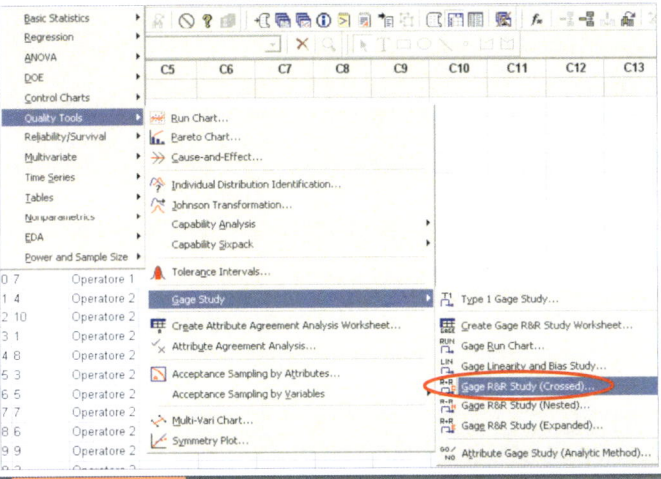

# Gage R&R (Continuous Data)

## Gage R&R Study (Crossed)…

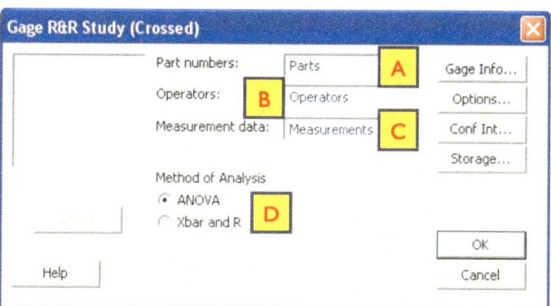

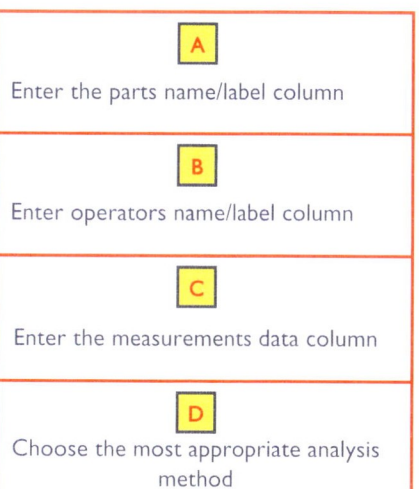

| | |
|---|---|
| **A** | Enter the parts name/label column |
| **B** | Enter operators name/label column |
| **C** | Enter the measurements data column |
| **D** | Choose the most appropriate analysis method |

# Gage R&R (Continuous Data)

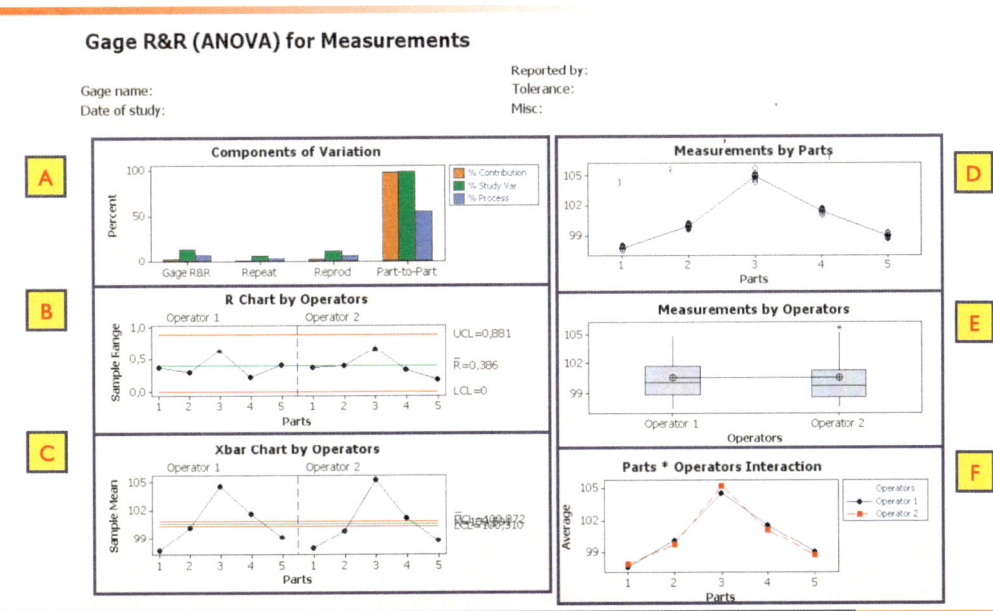

# Gage R&R (Continuous Data)

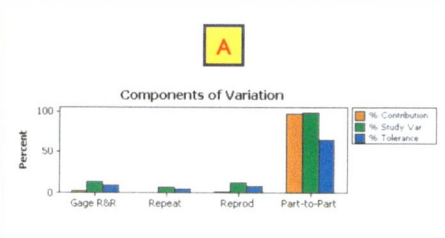

**A**

Components of Variation

- Each set of bars represents a source of variation. By default there are % *Contribution* and % *StudyVar*, but if you enter the option of *Tolerance* (or the *Historical Tolerance*) , it shows a third bar % *Tolerance*
- In a good measurement system the biggest contributor of variability should be the *Part-to-Part Variation*
- Note that, in case of % *StudyVar and % Tolerance*, the sum of the bar heights of repeatability and reproducibility will not be equal to the bar height of Gage R&R. This is because the standard deviations will not add up (unlike $\sigma^2$)

**B**

R Chart by Operators

- The *R-chart* shows the variability in the measurements for each operator
- Specifically, it shows:

  - each dot in the R-chart for a particular operator shows the range (maximum-minimum) of repeated measurements; if they are same, the range will be zero

  - the *Center line* is calculated by averaging all the ranges for each subgroup of repeated measurements for each part

  - *Control Limits* (UCL and LCL) are computed by using the variance within the subgroups

# Gage R&R (Continuous Data)

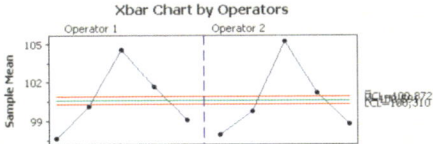

- The X-bar chart shows *Part-to-Part* variation by displaying the mean of repeated measurements of each part
- Specifically, it shows:
  - the dots in the chart represent, for each operator, the mean of repeated measurements of each part
  - the *Center Line* represents the mean of all measurements from all operators and all parts
  - *Control Limits* (UCL and LCL) are calculated from measurement data
- The measurement system is acceptable when plotted dots are out of control limits
- The variation caused by Repeatability should be much smaller than *Part-to-Part* variation

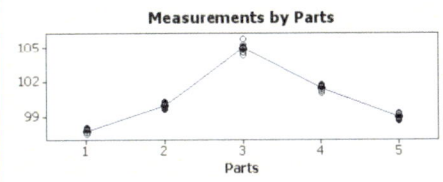

- This chart shows the dotted plots for the measured values of each part aggregating all operators
- In the chart, circles are individual measurements, solid dots represent mean value of measurements for each part
- Ideally:
  - measurements for each part should closely cluster together, with little variation
  - part-to-part measurement variation should be much larger than the variation of measurements on the same part

# Gage R&R (Continuous Data)

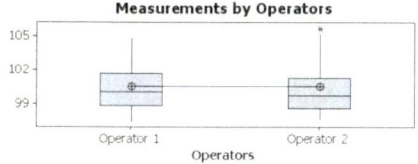

- *Measurement by operator* graph may help to determine if reproducibility will affect the measurements
- It shows all the measurements taken by different operators: the circles show averages. The line connects the mean measurements for each operator
- With this graph, we can also determine whether the total variability in the parts is the same for each operator (variability should have similar mean and variation)

## • Guidelines AIAG

| Line condition... | Indication... |
|---|---|
| Parallel to X-axis | Operators yield similar measurement means |
| Not parallel to X-axis | Operators yield different measurement means. This indicates operators measure in different ways |

# Gage R&R (Continuous Data)

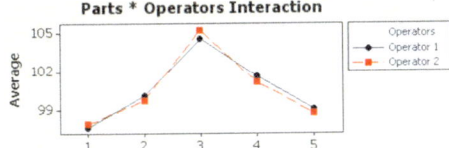

- This *Operator*Part Interaction* chart shows lines that connect the mean measurements for each operator measuring each part; each line indicates an operator
- In an ideal situation, all lines should be coincidental with each other, indicating no operator induced measurement variation among all parts

## • Guidelines AIAG

| Line condition | Indication |
| --- | --- |
| Virtually parallel | The operators measure parts in a similar way |
| A line is substantially higher or lower than others | An operator consistently yields measurements higher or lower than others |
| Non parallel or crossed lines | The results of the operators' measurements are affected by different parts |

# Gage R&R (Continuous Data)

## Two-Way ANOVA Table With Interaction

| Source | DF | SS | MS | F | P |
|---|---|---|---|---|---|
| Parts | 4 | 244,805 | 61,2012 | 129,630 | 0,000 |
| Operators | 1 | 0,022 | 0,0221 | 0,047 | 0,839 |
| Parts * Operators | 4 | 1,888 | 0,4721 | 13,973 | 0,000 |
| Repeatability | 30 | 1,014 | 0,0338 | | |
| Total | 39 | 247,729 | | | |

Alpha to remove interaction term = 0,25

## Gage R&R

| Source | VarComp | %Contribution (of VarComp) |
|---|---|---|
| Total Gage R&R | 0,14337 | 1,85 |
| Repeatability | 0,03379 | 0,44 |
| Reproducibility | 0,10958 | 1,42 |
| Operators | 0,00000 | 0,00 |
| Operators*Parts | 0,10958 | 1,42 |
| Part-To-Part | 7,59113 | 98,15 |
| Total Variation | 7,73451 | 100,00 |

| Source | StdDev (SD) | Study Var (6 * SD) | %Study Var (%SV) |
|---|---|---|---|
| Total Gage R&R | 0,37864 | 2,2719 | 13,61 |
| Repeatability | 0,18382 | 1,1029 | 6,61 |
| Reproducibility | 0,33103 | 1,9862 | 11,90 |
| Operators | 0,00000 | 0,0000 | 0,00 |
| Operators*Parts | 0,33103 | 1,9862 | 11,90 |
| Part-To-Part | 2,75520 | 16,5312 | 99,07 |
| Total Variation | 2,78110 | 16,6866 | 100,00 |

**Number of Distinct Categories = 10**

**A**

Identifying significance of source of variability by p value:

For a main effect, a p value less than 5% indicates its significance and for an interaction less than 25% may indicate its significance.

**B**

Rule of thumb: Performance level of a gage (threshold values of acceptability)

| % R&R (% Study Var) | Performance Level |
|---|---|
| 0 - 10 % | Good |
| 10 - 30 % | Marginal |
| > 30 % | Unacceptable |

**C**

NDC (*Number of Distinct Categories*) ≥ 5

LEAN SIX SIGMA MINIBOOK

# Gage R&R (Attribute Discrete Data)

Gage R&R Test Plan:

## Example of Gage R&R Test for Good/Not Good Judgments

| n°Parts | Standard | Group 1_Trial1 | Group 1_Trial2 | Group 2_Trial1 | Group 2_Trial2 | Group 3_Trial1 | Group 3_Trial2 |
|---------|----------|----------------|----------------|----------------|----------------|----------------|----------------|
| 1 | Not good | Not good | Not good | Not good | Not good | Not good | Not good |
| 2 | Not good | Not good | Not good | Not good | Not good | Not good | Not good |
| 3 | Not good | Not good | Not good | Not good | Not good | Not good | Not good |
| 4 | Good | Good | Good | Good | Good | Not good | Good |
| 5 | Not good | Good | Good | Good | Good | Not good | Not good |
| 6 | Not good | Good | Good | Good | Not good | Not good | Not good |
| 7 | Good | Good | Good | Not good | Good | Not good | Good |
| 8 | Good | Good | Good | Good | Good | Not good | Good |
| 9 | Not good | Not good | Not good | Not good | Not good | Not good | Not good |
| 10 | Good | Good | Good | Good | Good | Not good | Good |
| 11 | Not good | Good | Good | Good | Good | Not good | Good |
| 12 | Not good | Good | Good | Good | Not good | Not good | Not good |

- Select 20-30 parts that can show the variability of the process
- Use an expert to evaluate all selected parts in order to create a standard reference value
- Use 2 or 3 operators to evaluate all parts and give good/not good ratings
- Apply randomization and blinding in the test

# Gage R&R (Attribute Discrete Data)

MINITAB:

Stat > Quality Tools > Attribute Agreement Analysis…

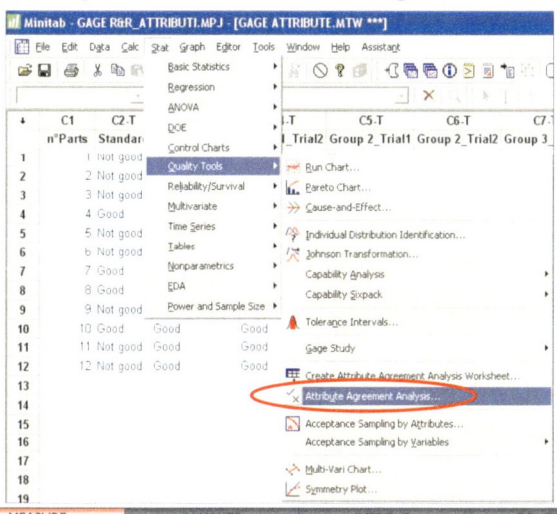

LEAN SIX SIGMA MINIBOOK

# Gage R&R (Attribute Discrete Data)

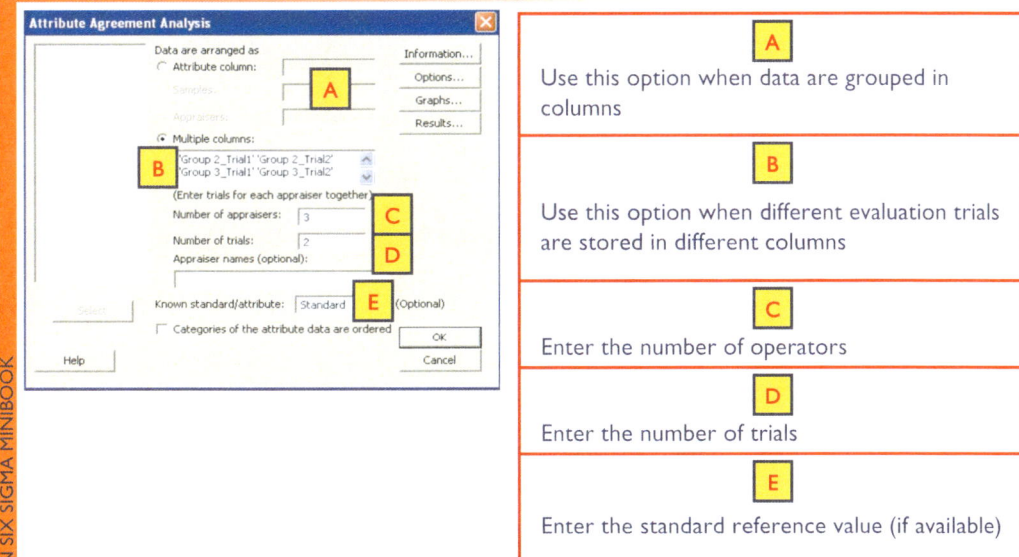

**A** — Use this option when data are grouped in columns

**B** — Use this option when different evaluation trials are stored in different columns

**C** — Enter the number of operators

**D** — Enter the number of trials

**E** — Enter the standard reference value (if available)

# Gage R&R (Attribute Discrete Data)

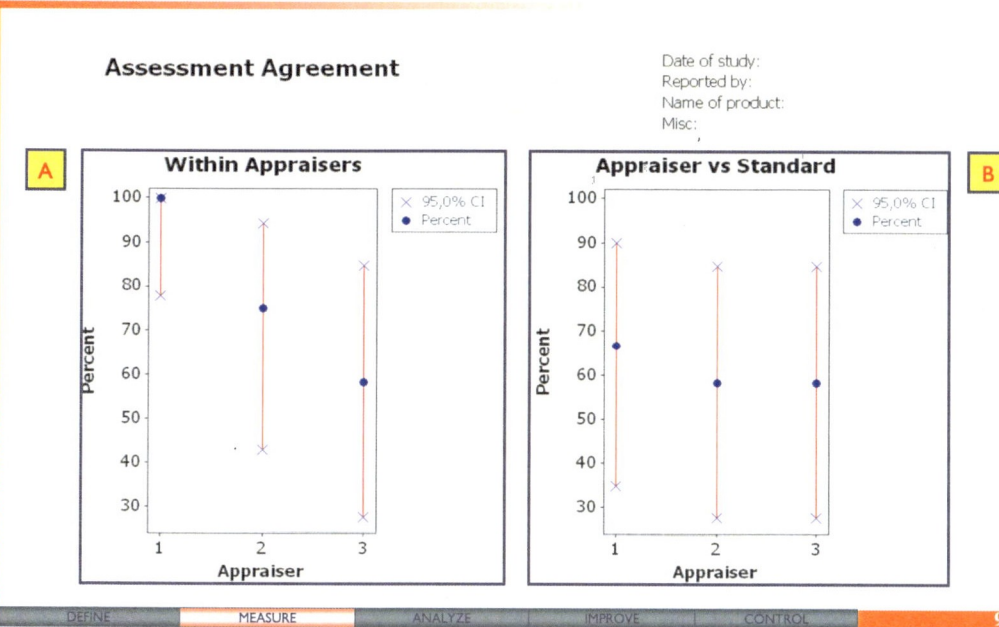

# Gage R&R (Attribute Discrete Data)

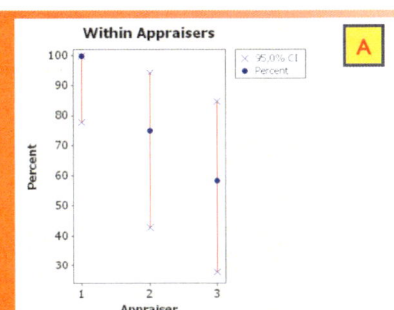

A

- This chart shows the consistency in responses for each operator
- For each operator
  - the blue dot shows the percentage of consistency in his/her answers
  - the red line gives the 95% confidence interval for the percentage of consistency in his/her answers
  - the blue X provides the upper and lower confidence limits

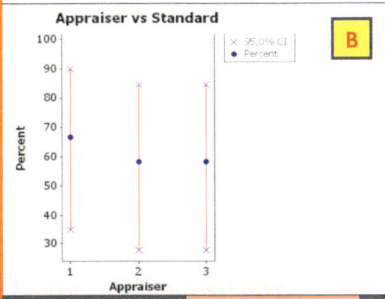

B

- This chart is generated when there is a standard as reference
- This chart shows the correctness of responses for each operator (accuracy in the answer)
- For each operator, the chart provides the following information:
  - the blue dot shows the percentage of consistency with the standard of reference
  - The red line represents 95% confidence interval of the percentage of consistency with the standard of reference
  - The blue X represents the upper and lower limit points of the confidence interval

LEAN SIX SIGMA MINIBOOK

# Gage R&R (Attribute Discrete Data)

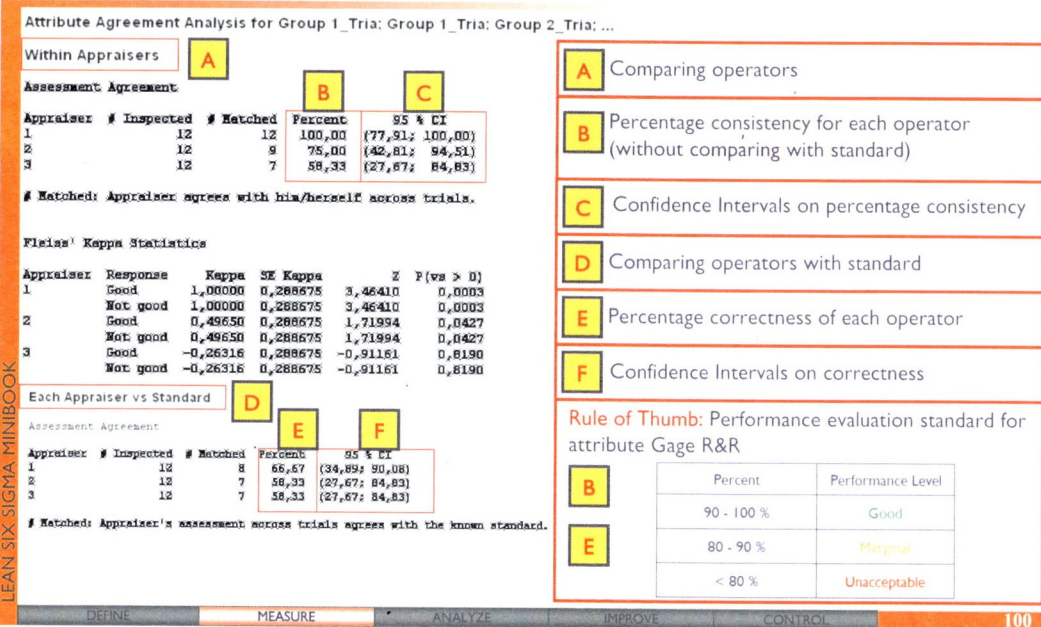

LEAN SIX SIGMA MINIBOOK

# Gage R&R (Attribute Discrete Data)

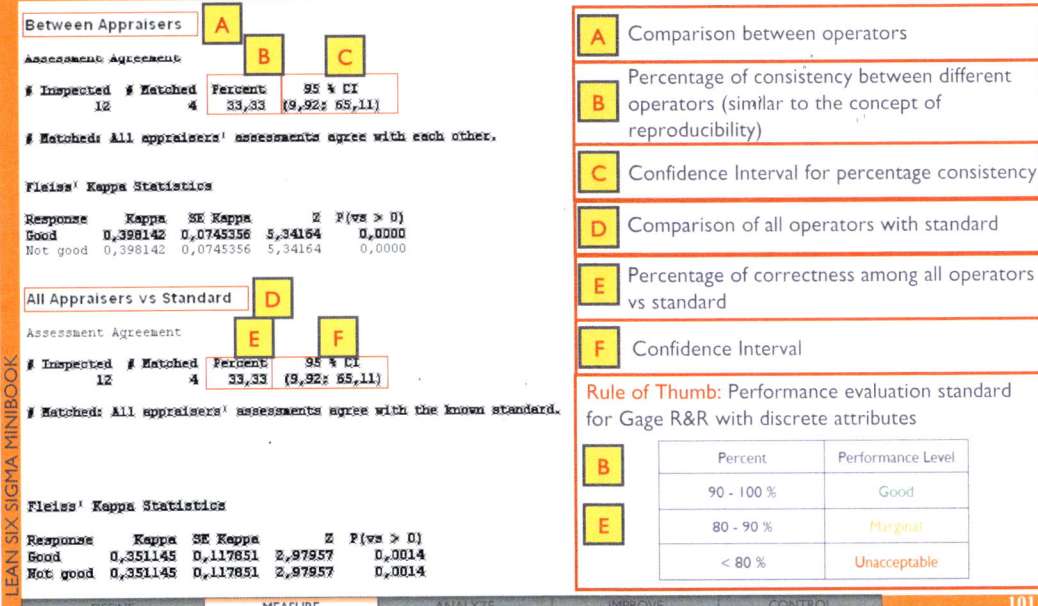

# Gage R&R & Minitab

Minitab Assistant helps you to choose the right Gage R&R methodology

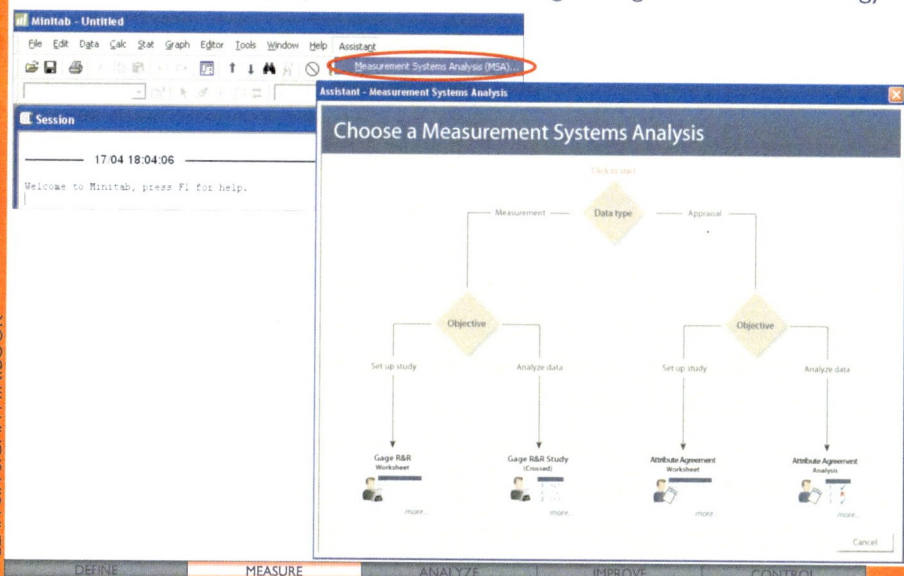

# Pareto Diagram

## Objective:

- Pareto diagram can help to identify high priority actions or areas, by graphically displaying them in terms of frequencies or scores in decreasing order, which allows us to focus our intervention and resources in key areas

## Features:

- This tool will allow management to focus on high impact area
- This is also a supporting tool for diagnosis of current situation and determination of priorities

# Pareto Diagram

MINITAB:

## Stat > Quality Tools > Pareto Chart...

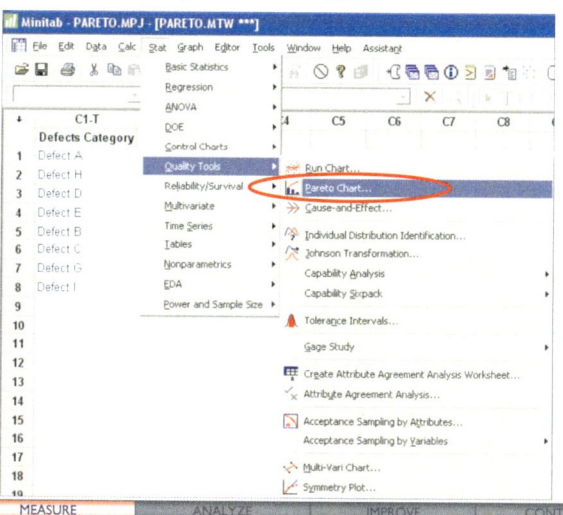

# Pareto Diagram

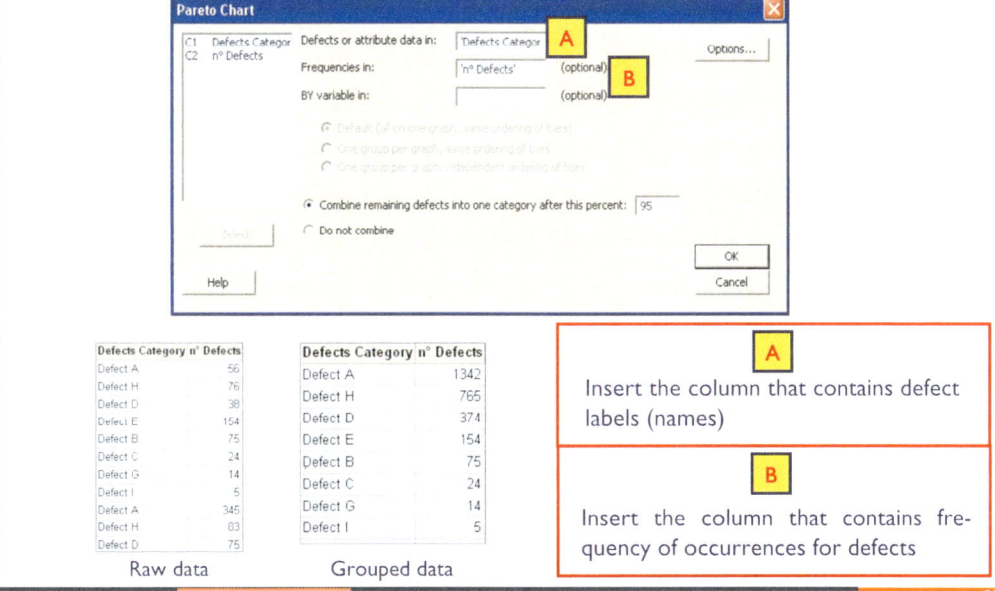

LEAN SIX SIGMA MINIBOOK

# Pareto Diagram

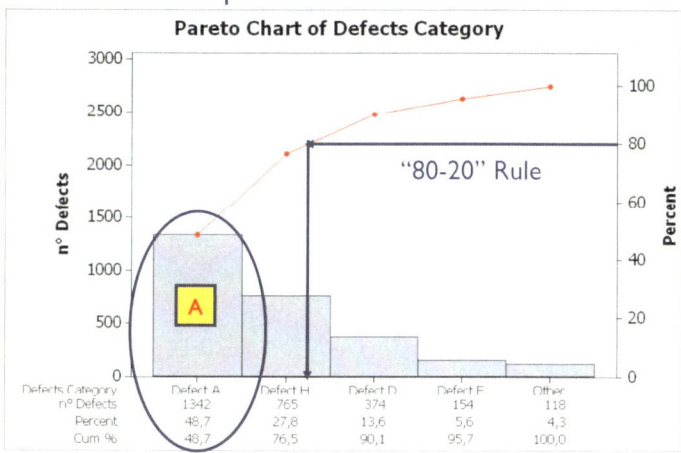

**Pareto Chart of Defects Category**

A is the item that is most critical to business:

Completely removing defect type A would reduce total defects by 48.7%

LEAN SIX SIGMA MINIBOOK

# Normality Test

Objective:

- The normality test is a statistical test that checks the validity of the assumption on data to be normally distributed. This normality assumption is critical for many statistical analyses

Statistical tools that need normality assumption:

- *1-Sample t*
- *2-Sample t*
- *Paired t-Test*
- ANOVA
- *Control Chart* for continuous data
- *Capability Analysis*

# Normality Test

## MINITAB:

### Stat > Basic Statistics > Normality Test...

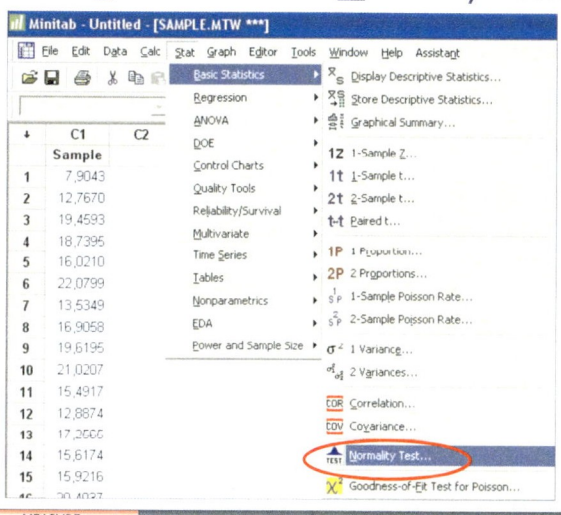

LEAN SIX SIGMA MINIBOOK

# Normality Test

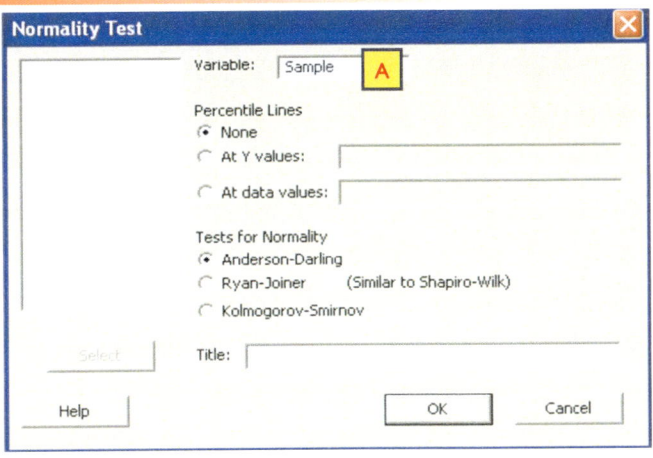

Insert the column that contains data to be tested for normality

# Normality Test

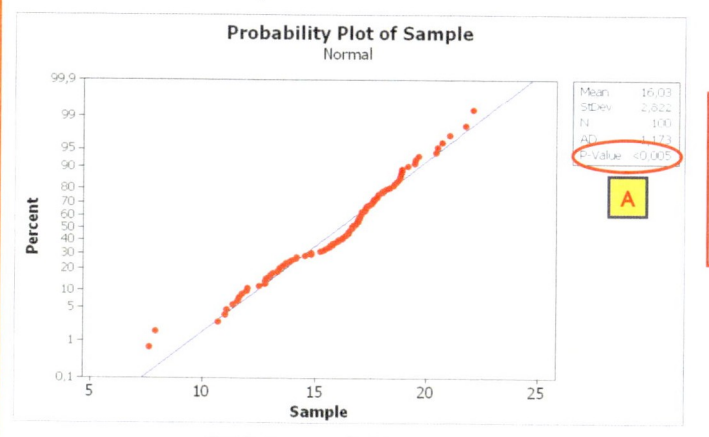

## MINITAB: Output

**Probability Plot of Sample**
Normal

Mean 16,03
StDev 2,822
N 100
AD 1,173
P-Value <0,005

Data are **not normal** because the P-Value is less than **0.05**

P-Value > 0.05 ⟶ Data are normal

P-Value < 0.05 ⟶ Data are not normal

LEAN SIX SIGMA MINIBOOK

# Capability Analysis

## Objective:

- Capability Analysis or *Process Capability Analysis* (G) is a study to determine the ability of current process to satisfy customer required specifications

## Basic Assumptions:

- Normal data
- Stable process

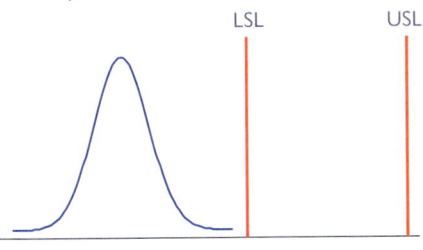
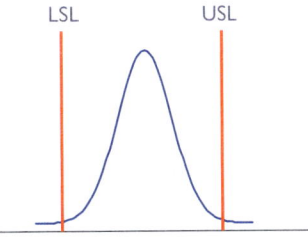

# Capability Analysis

| Short Term (ST) | |
|---|---|
| Potential capability index | Actual capability index |
| **A** $C_p = \dfrac{USL - LSL}{6\sigma_{ST}}$ | **B** $C_{pk} = \min\left\{ \dfrac{USL - \mu}{3\sigma_{ST}}; \dfrac{\mu - LSL}{3\sigma_{ST}} \right\}$ |

| | | |
|---|---|---|
| **A** | 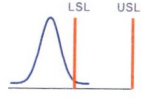 | Cp does not take into account the process centering |
| **B** | 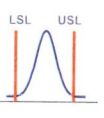 | Cpk takes into account the process centering |

# Capability Analysis

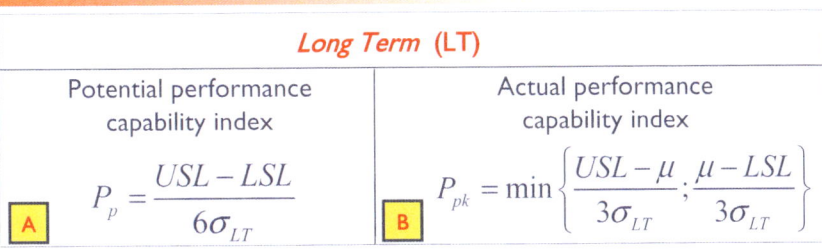

| | Long Term (LT) | |
|---|---|---|
| **Potential performance capability index** | **Actual performance capability index** | |
| **A** $P_p = \dfrac{USL - LSL}{6\sigma_{LT}}$ | **B** $P_{pk} = \min\left\{\dfrac{USL - \mu}{3\sigma_{LT}}; \dfrac{\mu - LSL}{3\sigma_{LT}}\right\}$ | |

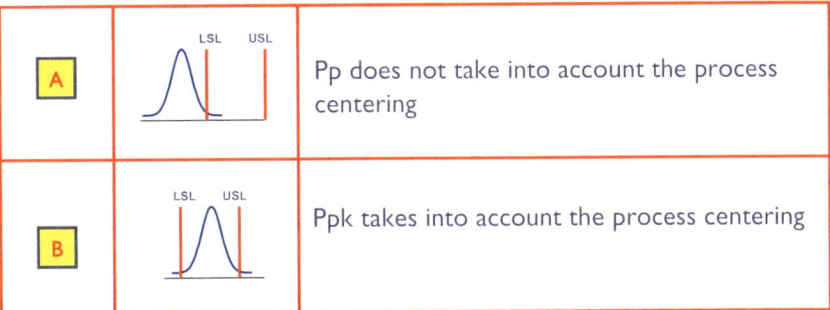

| **A** | LSL USL | Pp does not take into account the process centering |
|---|---|---|
| **B** | LSL USL | Ppk takes into account the process centering |

# Capability Analysis

MINITAB:

Stat > Quality Tools > Capability Sixpack > Normal…

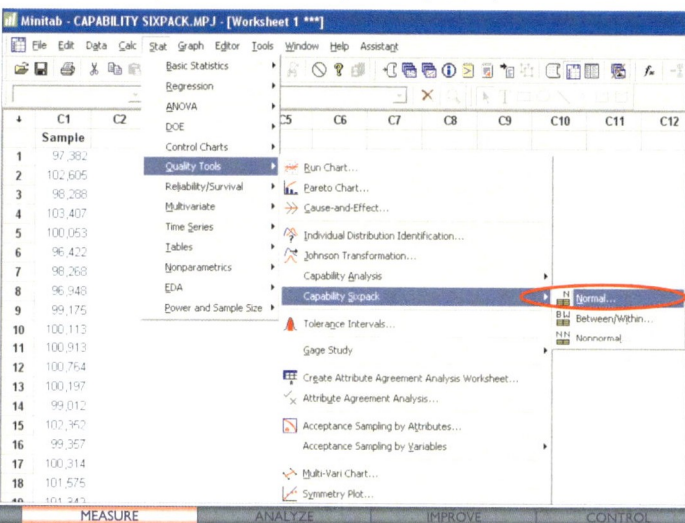

# Capability Analysis

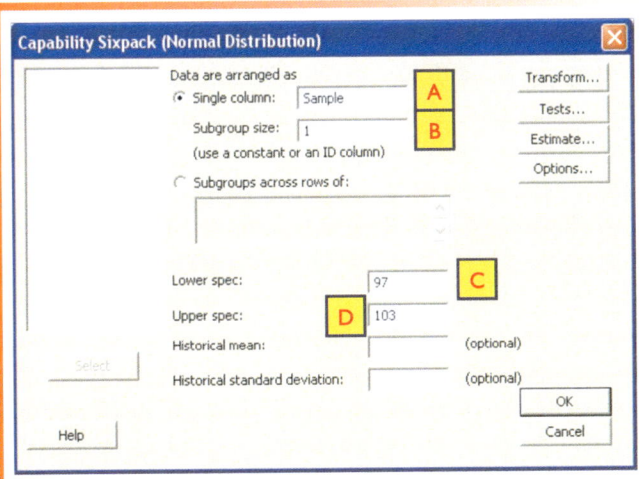

**Capability Sixpack (Normal Distribution)**

Data are arranged as

⊙ Single column: | Sample |  **A**

Subgroup size: | 1 | **B**

(use a constant or an ID column)

◯ Subgroups across rows of:

Lower spec: | 97 | **C**

Upper spec: | **D**  103 |

Historical mean: | | (optional)

Historical standard deviation: | | (optional)

Transform...
Tests...
Estimate...
Options...

Select

Help

OK
Cancel

**A**

Insert the column that contains process data to be assessed for capability

**B**

Insert subgroup size by either inputing a number or a subgroup index column

**C**

Insert lower specification limit

**D**

Insert upper specification limit

LEAN SIX SIGMA MINIBOOK

# Capability Analysis

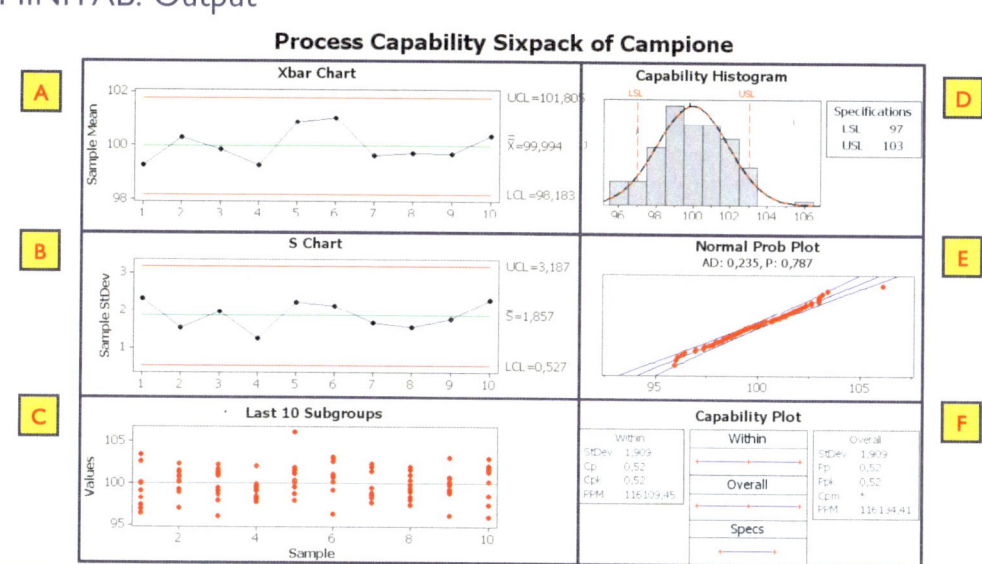

Process Capability Sixpack of Campione

# Capability Analysis

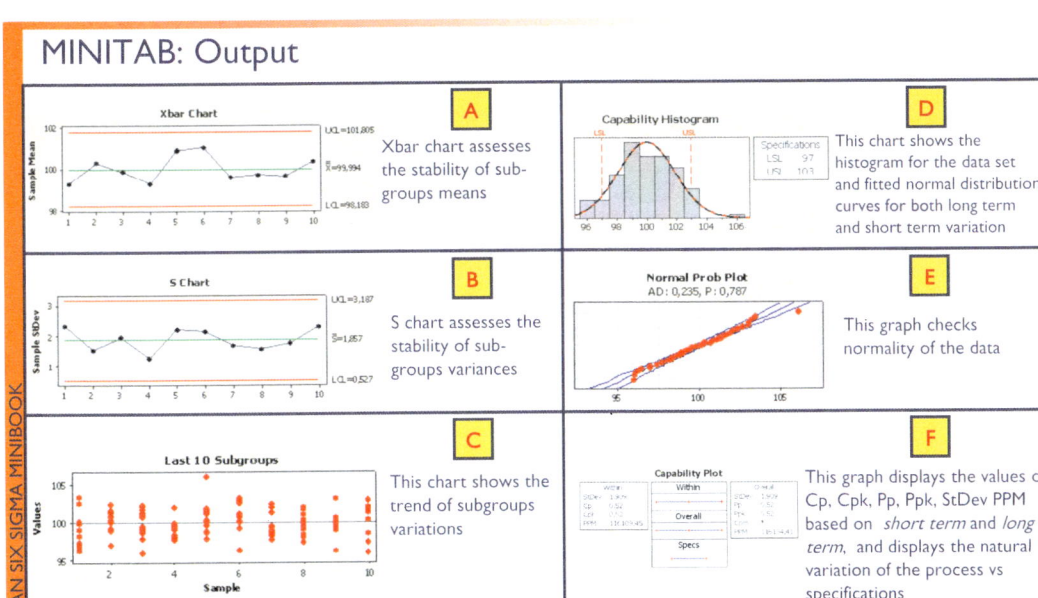

## MINITAB: Output

**A**
Xbar chart assesses the stability of sub-groups means

**B**
S chart assesses the stability of sub-groups variances

**C**
This chart shows the trend of subgroups variations

**D**
This chart shows the histogram for the data set and fitted normal distribution curves for both long term and short term variation

**E**
This graph checks normality of the data

**F**
This graph displays the values of Cp, Cpk, Pp, Ppk, StDev PPM based on *short term* and *long term*, and displays the natural variation of the process vs specifications

# Calculation of DPMO

## Objective:

- DPMO[G], *i.e. Defects Per Million of Opportunity* is a performance indicator calculated as a ratio of number of defects divided by the maximum number of potential defects in a batch of units inspected

## Definitions:

- U = number of units inspected
- D = number of total defects
- O = number of opportunities for defect per unit inspected. It is the maximum number of potential defects of all failure modes for a unit

$$DPO = \frac{D}{U \times O} \implies DPMO = \frac{D}{U \times O} \times 1000000$$

# Calculation of Process Sigma

Objective:
- *Process Sigma* (G) calculates Sigma level of a process based on defects detected in a batch of units

Calculation procedure:

$$DPO = \frac{51}{1000 \times 2}$$

$$\eta = 1 - \frac{51}{1000 \times 2} = 0.9745$$

|  | |  |
|------|---|-------|
| Step 1 | Number of opportunity per unit (O) | 2 |
| Step 2 | Number of units inspected (U) | 1000 |
| Step 3 | Number of defects found (D) | 51 |
| Step 4 | Number of defects divided by total number of opportunities | 0,0255 |
| Step 5 | Calculate performance level (process yield) | 0,9745 |
| Step 6 | Determine Process Sigma value | 3,45 |

| Yield | DPMO | Sigma |
|-------|------|-------|
| 95,05285% | 49471 | 3,15 |
| 95,54345% | 44565 | 3,20 |
| 95,99408% | 40059 | 3,25 |
| 96,40697% | 35930 | 3,30 |
| 96,78432% | 32157 | 3,35 |
| 97,12834% | 28717 | 3,40 |
| 97,44119% | 25588 | 3,45 |
| 97,72499% | 22750 | 3,50 |
| 97,98178% | 20182 | 3,55 |

# Takt Time

## Objective:

- The *Takt Time* (G) represents the rhythm of production/delivery that a process (workstation, Cell, etc.) must respect to satisfy customer demand. Each step of the process must produce to Takt Time to ensure a stable flow of outputs to meet customer requirements

## Calculation method:

$$\text{TAKT TIME} = \frac{\text{Available Time (Time)}}{\text{Customer Demand (pcs/Time)}}$$

**Available Time**: Total available time minus planned downtime (example: Breaks)

**Customer Demand**: Total expected demand from Customer (pieces per unit of time)

$$\text{N}° \text{ Operators necessary} = \frac{\text{Cycle Time}}{\text{Takt Time}}$$

**Cycle Time**: Total manual working time for one cycle (Touch Time) plus automatic run time. In case of production mix, the weighted mean (weighted on quantity) can be useful

# Takt Time

**Step n° 1:** Determine how fast the process (Takt Time) must be

- Customer Demand: 35000 units/year
- Time per shift: 8 hours (1 shift per day)
- Breaks (2 x 15 minutes): 30 minutes

Time per shift (8 hours x 3600 seconds) = 28800 s
Breaks(30 minutes x 60 seconds) = 1800 s
**Available time**= 27000 s
**Customer Demand / day** = 35000 / 220 = 159 pieces

$$\text{TAKT TIME} = \frac{27000}{159} = 170 \text{ s}$$

The customer buys goods at the rate of one every 170 seconds. This is the "Target" frequency which all processes must reach to produce each part or component to meet customer demand

**Step n° 2:** Identify how many people are necessary for each process to satisfy Customer demand

Traditional production line

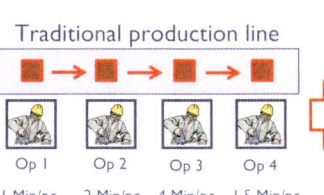

Op 1    Op 2    Op 3    Op 4
1 Min/pc   2 Min/pc   4 Min/pc   1,5 Min/pc

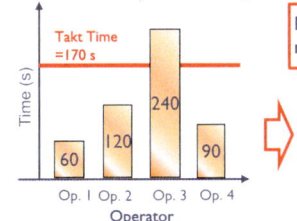

- **Cycle time** = 60 + 120 + 240 + 90 = 510 s
(total time necessary for the production of one piece)

$$\text{No. Operators necessary} = \frac{510}{170} = 3 \text{ Ops.}$$

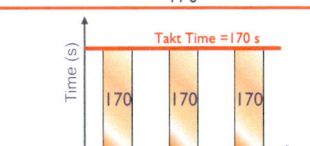

**New asset:** the contribution of the three operators in the production of each piece is the same and equal to Takt Time

# Overall Equipment Effectiveness (OEE)

## Objective:

- The OEE[G] (Overall Equipment Effectiveness) is a powerful method to monitor and improve the efficiency of manufacturing and transactional processes. OEE is frequently used as a key metric in TPM (Total Productive Maintenance) and Lean programs. One of the main goals is to reduce what are called the *Six Big Losses* clustered in three categories: **Downtime**, **Speed** and **Quality losses**

| SIX BIG LOSSES | CATEGORY | EXAMPLE | NOTES |
|---|---|---|---|
| Breakdowns | Downtime Loss | Equipment failure<br>Unplanned maintenance | The threshold value between a breakdown and a minor stoppage can change from company to company (in most common situations 5 minutes) |
| Setup | Downtime Loss | Setup/Changeover<br>Material/Operator shortages<br>Warm up Time | A good way to reduce this kind of loss is SMED (set-up reduction methodology) |
| Minor Stoppages | Speed Loss | Parts jam<br>Checking<br>Framed pieces | The general definition of minor stoppages is a stop less than 5 minutes that does not require Maintenance operator intervention |
| Reduced Speed | Speed Loss | Equipment wear<br>Lack of equipment knowledge<br>Operator inefficiency | These kinds of causes prevent the process from operating at maximun speed (theoretical speed) |
| Start Up Losses | Quality Loss | Scraps<br>Reworks | Scraps produced during start up phase |
| Scraps and reworks | Quality Loss | Scraps<br>Reworks | Scraps or reworks produced during production phase |

# Overall Equipment Effectiveness (OEE)

## Overview:

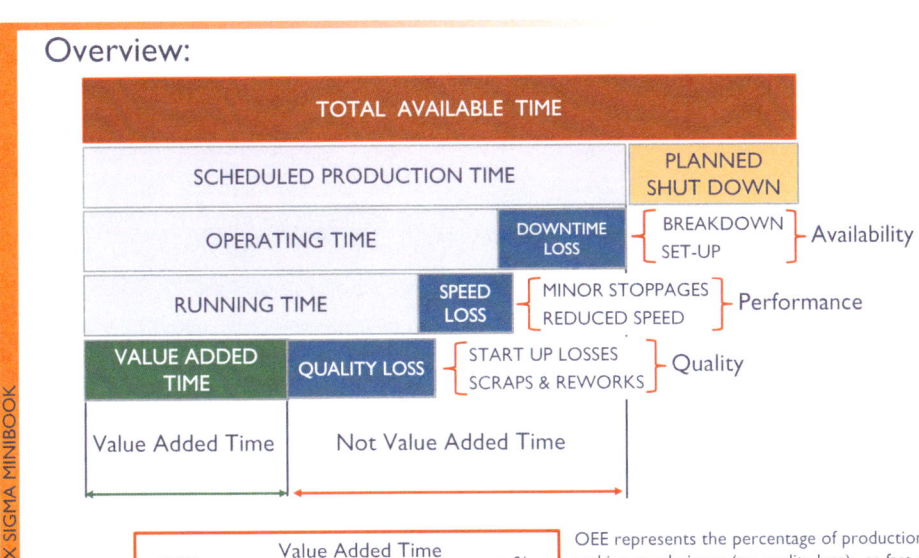

$$OEE = \frac{Value\ Added\ Time}{Scheduled\ production\ Time} = \%$$

OEE represents the percentage of production time spent making good pieces (no quality loss), as fast as possible (no speed loss), without interruption (no downtime)

LEAN SIX SIGMA MINIBOOK

# Overall Equipment Effectiveness (OEE)

## How to calculate OEE index:

$$OEE = \frac{\text{Number of acceptable units produced}}{\text{Theoretical number of units that could have been produced in the scheduled time running at standard speed}} \times 100\%$$

**Formula based on units**

- Standard speed = 110 pieces/minutes

- Number of good units produced = 11253 pieces

| SET UP | PKG | CLEANING | PKG | MINOR STOP | PKG | CHARGE MATERIAL | PKG |
|--------|-----|----------|-----|------------|-----|-----------------|-----|

Total time = 280 minutes

$$OEE = \frac{11253}{110 \times 280} \times 100\% = \frac{11253}{30800} \times 100\% = 36\%$$

Average OEE = 60%
World Class OEE = 85%

LEAN SIX SIGMA MINIBOOK

# Overall Equipment Effectiveness (OEE)

How to calculate OEE index considering the three OEE factors:

$$OEE = \text{Availability} \times \text{Performance} \times \text{Quality} \times 100\%$$

$$\text{Availability} = \frac{\text{Operating Time}}{\text{Scheduled Production Time}} \times 100\%$$

World Class = 90%

$$\text{Performance} = \frac{\text{Total pieces} \times \text{Theoretical time to produce 1 unit}}{\text{Operating time}} \times 100\%$$

World Class = 95%

$$\text{Quality} = \frac{\text{Good pieces}}{\text{Total pieces}} \times 100\%$$

World Class = 99,9%

$$OEE = \frac{\text{Number of good pieces produced} \times \text{Theoretical time to produce 1 unit}}{\text{Scheduled production time}} \times 100\%$$

Formula based on time

# Overall Equipment Effectiveness (OEE)

## Example of the OEE data collection:

# Overall Equipment Effectiveness (OEE)

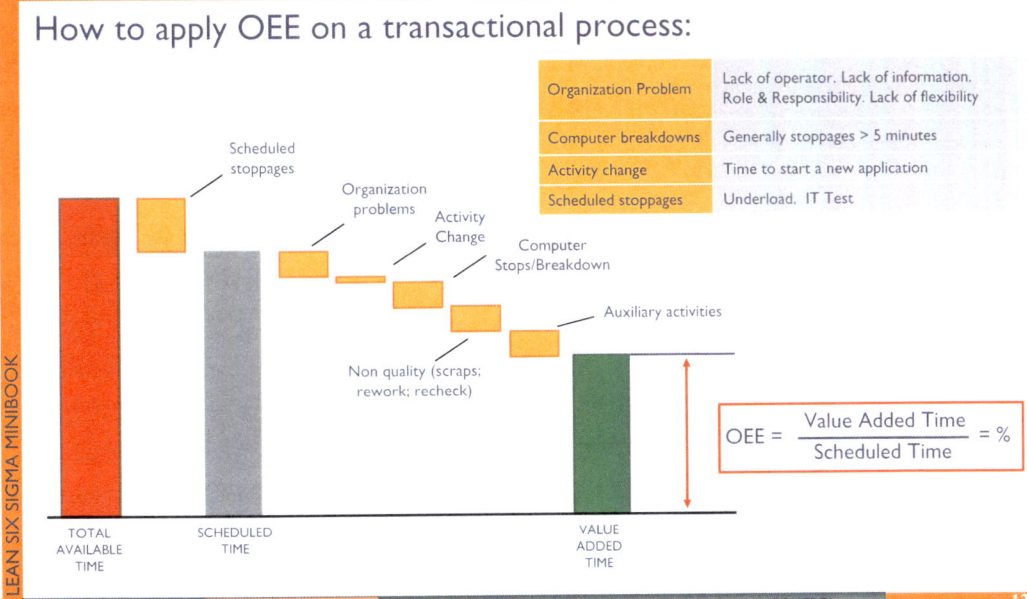

How to apply OEE on a transactional process:

| | |
|---|---|
| Organization Problem | Lack of operator. Lack of information. Role & Responsibility. Lack of flexibility |
| Computer breakdowns | Generally stoppages > 5 minutes |
| Activity change | Time to start a new application |
| Scheduled stoppages | Underload. IT Test |

Scheduled stoppages

Organization problems

Activity Change

Computer Stops/Breakdown

Auxiliary activities

Non quality (scraps; rework; recheck)

$$OEE = \frac{Value\ Added\ Time}{Scheduled\ Time} = \%$$

TOTAL AVAILABLE TIME

SCHEDULED TIME

VALUE ADDED TIME

# Time Series Plot

Objective:

- Time Series Plot is a tool to analyze a potential time trend in a sequence of data

Features:

- It is a tool to analyze any trends over time, to assess the need to stratify the data (eg. cyclic patterns), to explore and validate the data before applying other tools (eg. Regression modeling)

# Time Series Plot

Stat > Time Series > Time Series Plot...

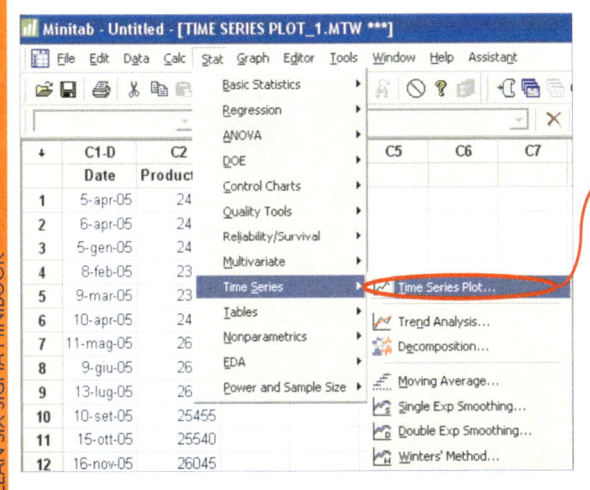

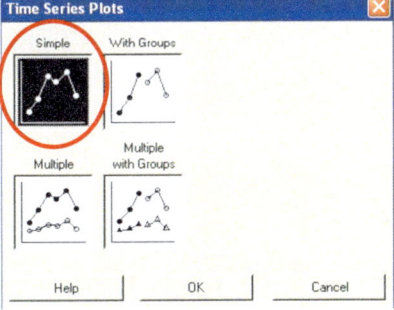

# Time Series Plot

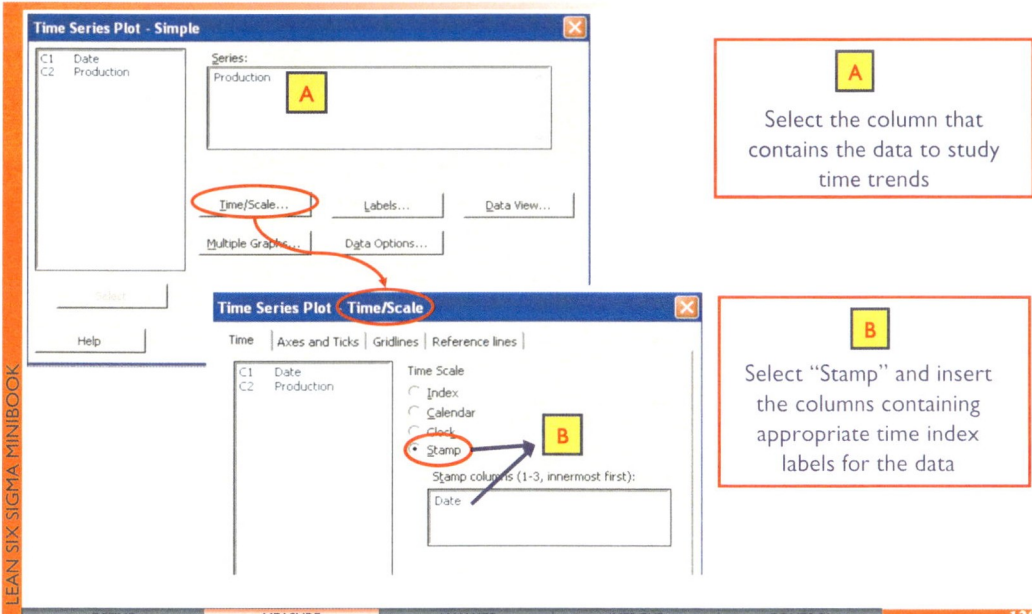

**A**

Select the column that contains the data to study time trends

**B**

Select "Stamp" and insert the columns containing appropriate time index labels for the data

# Time Series Plot

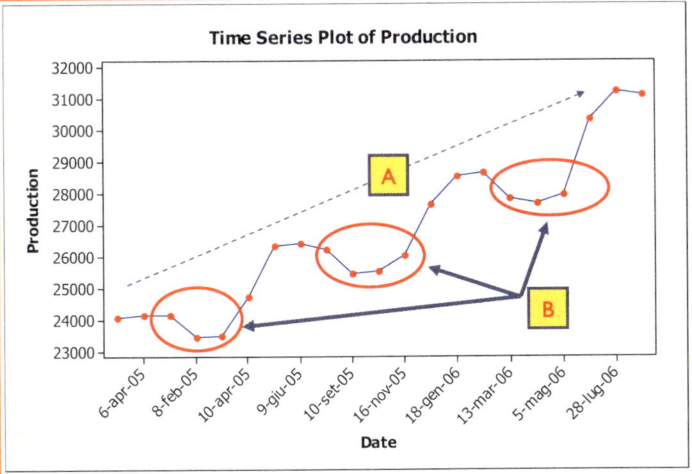

Time Series Plot of Production
Production
32000
31000
30000
29000
28000
27000
26000
25000
24000
23000
A
B
6-apr-05
8-feb-05
10-apr-05
9-giu-05
10-set-05
16-nov-05
18-gen-06
13-mar-06
5-mag-06
28-lug-06
Date


**A**

It is possible to observe an increasing trend for production over time

**B**

Possible cyclic pattern: stratification might be needed (eg. day of the week)

# Run Chart

## Objective:

- Run Chart is a tool that identifies possible special causes for the process performance variation

## Features:

- **Common Causes**(G): Random causes are not attributable to special events, but to natural variability inherent in all processes

- **Special Causes**(G): Special causes may be associated with special events, or else the result of temporal trends (*Clustering, Mixture, Trend, Oscillation*). If they are present the process will be "out of control"

---

### CAUTION:

The use of a Run Chart doesn't require normality assumption in process data

---

LEAN SIX SIGMA MINIBOOK

# Run Chart

**Patterns that indicate special causes:**

- *Clustering* is a pattern characterized by grouped data in a certain area of the graph (Clustering could be caused by variation in measurement systems, batch-batch variation, sampling issues)
- *Mixture* is a pattern in which a few points are near the median (Mixture is usually caused by a combination of two populations or processes operating at different levels)
- *Trend* is a pattern in a sequence of points increasing or decreasing (Trend could be caused by such things as worn out tools and fatigue of workers)
- *Oscillation* indicates fluctuating data that moves up and down rapidly and could mean that the process is unstable

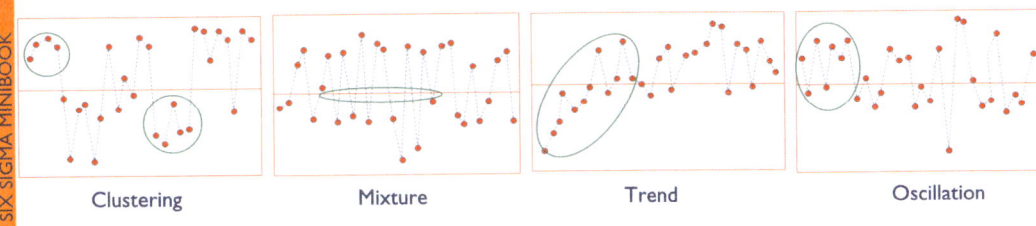

| Clustering | Mixture | Trend | Oscillation |

Run is defined as the number of consecutive points that are on the same side than the median

# Run Chart

MINITAB:

## $\underline{S}$tat > $\underline{Q}$uality Tools > $\underline{R}$un Chart…

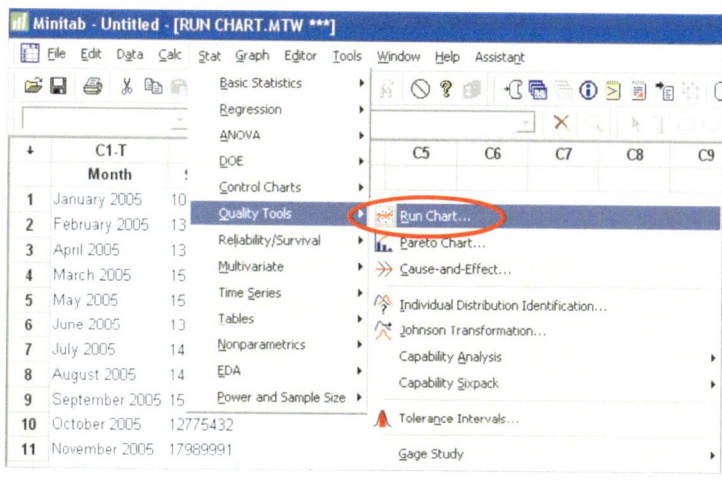

DEFINE    MEASURE    ANALYZE    IMPROVE    CONTROL

# Run Chart

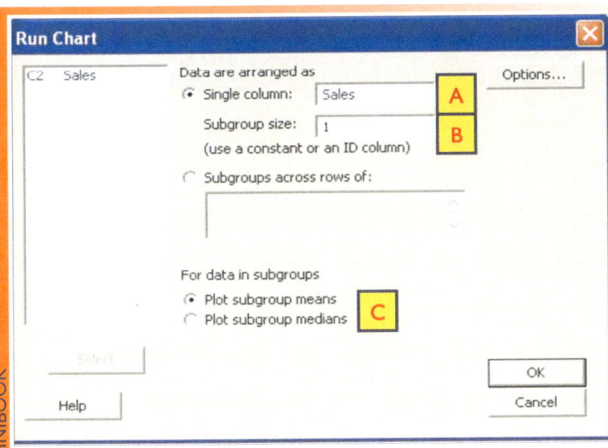

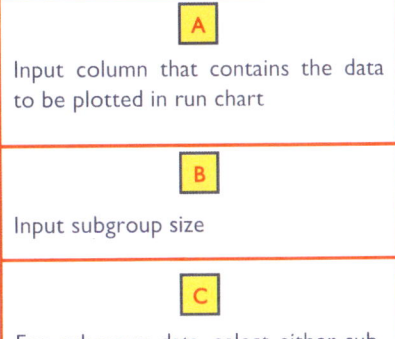

**A**

Input column that contains the data to be plotted in run chart

**B**

Input subgroup size

**C**

For subgroup data, select either subgroup mean or median to be plotted in order to track the change in central tendency

# Run Chart

**Run Chart of Sales**

| Number of runs about median: | 11 | Number of runs up or down: | 19 |
| Expected number of runs: | 13,0 | Expected number of runs: | 15,7 |
| Longest run about median: | 6 | Longest run up or down: | 2 |
| Approx P-Value for Clustering: | 0,202 | Approx P-Value for Trends: | 0,953 |
| Approx P-Value for Mixtures: | 0,798 | Approx P-Value for Oscillation: | 0,047 | A |

**A**

The P-Value is less than 0.05 and indicates that there is significant oscillation

# ANALYZE

Arcidiacono G., Calabrese C., Yang K.: Leading processes to lead companies: Lean Six Sigma.
DOI 10-1007/978-88-470-2492-2, © Springer-Verlag Italia 2012

# ANALYZE

*Analyze* is the third step in a Lean Six Sigma project roadmap. In this phase we try to:

- Explore the relationships among variables and start root cause analysis of major problems
- Conduct cause - effect analysis for trouble shooting
- Discover the real root causes rather than take care of symptoms
- Use statistical significance testing as a tool to identify key variables for response

# Cause-Effect Diagram

Objective:

- The Cause-Effect Diagram is a visual tool that can help to identify the relationship between an **effect** and its possible **root causes**

- The Cause-Effect Diagram is also an effective tool for quality management and Brainstorming[G]. It is one of the common tools for Lean Six Sigma projects and problem solving techniques

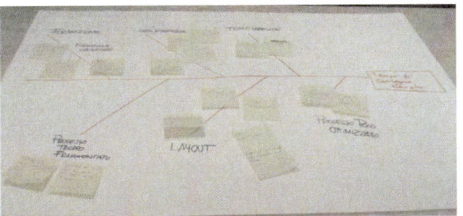

# Cause-Effect Diagram

## Why use the Cause-Effect Diagram?

- It is easy to understand and is a comprehensive graphic template that can logically display complex cause and effect relationships
- It can improve the understanding of a detailed relationship between cause and effect

## When to use it:

- If there is a large number of possible root causes
- When the relationship between cause and effect is not clear

# Cause-Effect Diagram

How to build a Cause-Effect Diagram:

1. Define the problem and identify the effect to be analyzed
2. Identify the categories of possible root causes (commonly used categories: *Measurements, Machines, Man/Personnel, Materials, Methods, Mother Nature/Environment*)
3. Identify potential causes and group them into categories. The method of the *5 Whys* can be used to determine the exact relationship between causes and effect
4. Sort the causes according to the possible degree of influence towards the effect

# Cause-Effect Diagram

MINITAB:

Stat > Quality Tools > Cause-and-Effect…

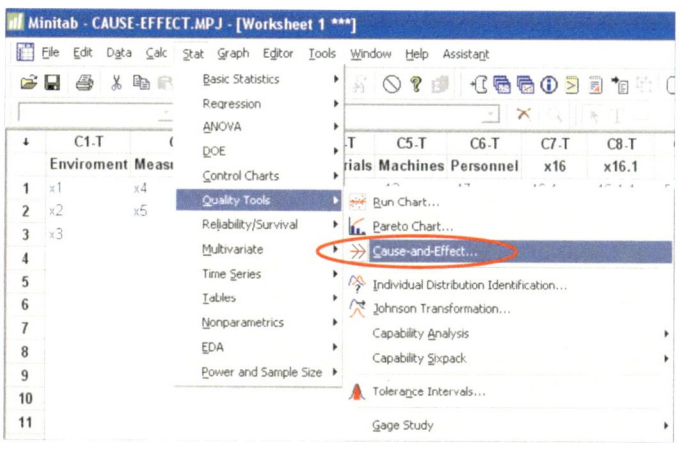

# Cause-Effect Diagram

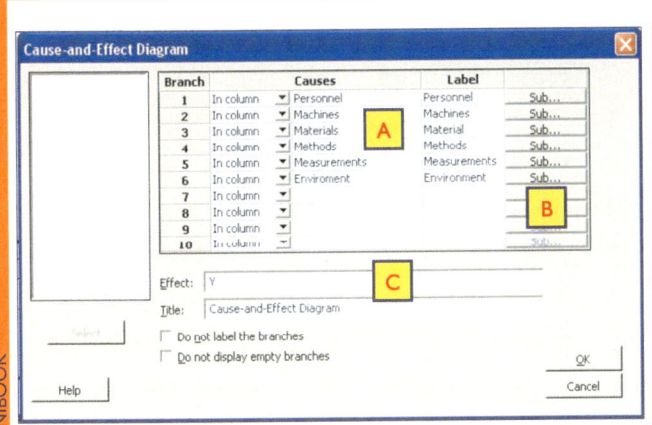

Enter the columns that contain the main categories of causes

Enter the columns that contain the secondary causes

Enter the effect to be analyzed

# Cause-Effect Diagram

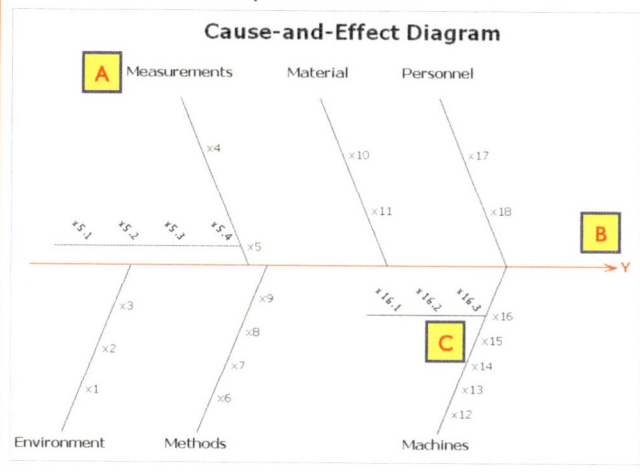

## MINITAB: Output

**Cause-and-Effect Diagram**

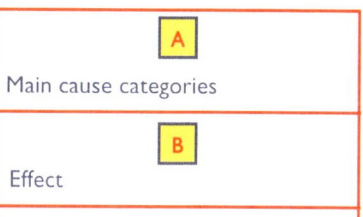

**A**
Main cause categories

**B**
Effect

**C**
Root causes drill down by 5 whys:
- **Why** Y happens? Because of Machine
- **Why** machine influences Y? Because X16
- **Why** X16 influences machine? Because X16.1 X16.2 e X16.3
- **Why**…

# Statistical Hypothesis Testing

Objective:

- Statistical Hypothesis Testing is used to make an inference or conclusion for a population, starting from sample data observation

- Typical applications of Hypothesis Testing are:

    - the comparison of means between two or more groups
    - the comparison of variances between two or more groups
    - the comparison of proportions, also extracted from samples of different sizes

# Statistical Hypothesis Testing

There are many types of hypothesis testing. We need to select the right Hypothesis Testing method for the right problem, as illustrated by the following table:

| | HYPHOTESIS TESTING | PURPOSE |
|---|---|---|
| **Mean comparison** | 1-Sample t | To compare means between a sample and a reference known mean |
| | 2-Sample t | To compare means between two groups |
| | Paired t-TEST | To compare means between two groups when data are paired |
| | ANOVA (F TEST) | To compare means between more than two groups |
| **Variance comparison** | Variance Test | To compare variances between two or more groups |
| **Proportion comparison** | Chi- Square Test | To compare proportions between two or more groups |

# Hypothesis Testing: 1-Sample t

Objective:

- The 1-Sample t test compares the mean of a sample with a given value

Fundamental Assumptions:

- Data should be normally distributed

How to read the test result:

- It is based on P-Value

| | |
|---|---|
| **P-Value > 0.05** | There is no significant difference between the population mean, from which the sample comes, and the given value |
| **P-Value ≤ 0.05** | The population mean is significantly different than the given value |

# Hypothesis Testing: 1-Sample t

MINITAB:

<u>S</u>tat > <u>B</u>asic Statistic > <u>1</u>-Sample t...

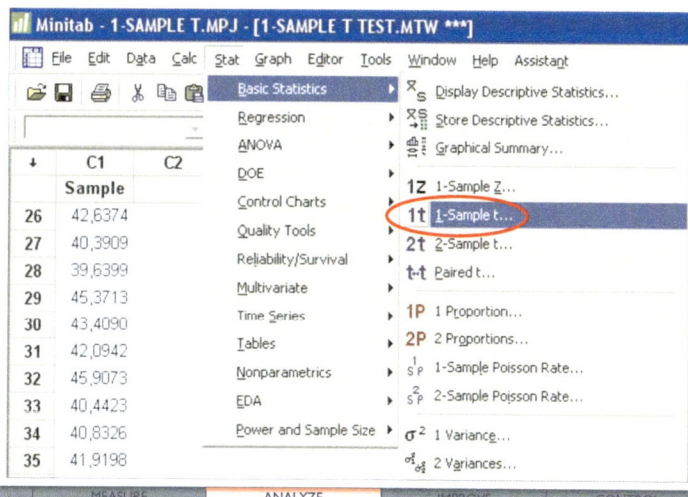

# Hypothesis Testing: 1-Sample t

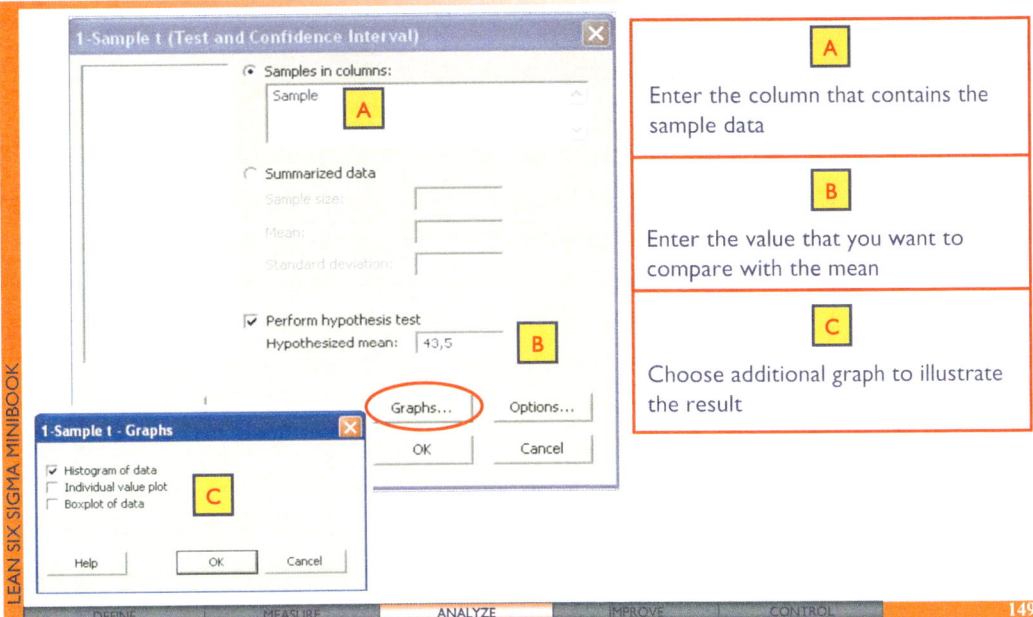

A — Enter the column that contains the sample data

B — Enter the value that you want to compare with the mean

C — Choose additional graph to illustrate the result

# Hypothesis Testing: 1-Sample t

## MINITAB: Output

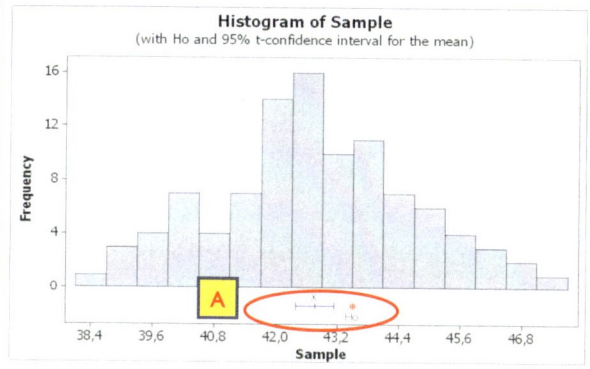

**Histogram of Sample**
(with Ho and 95% t-confidence interval for the mean)

**A**

You can graphically observe if the Confidence Interval for the population mean contains the given reference value, if the interval doesn't contain the value, it indicates that population mean is not equal to the given value

**B**

With P-Value ≤ 0.05, it indicates that the population mean is statistically and significantly different than the given value

```
One-Sample T: Sample

Test of mu = 43,5 vs not = 43,5


Variable    N    Mean   StDev  SE Mean     95% CI              T      P
Sample     100  42,762  1,901   0,190   (42,384; 43,139)    -3,88  0,000
```

# Hypothesis Testing: 2-Sample t

**Objective:**

- The 2-Sample t test compares the **means** of 2 populations

**Fundamental Assumption:**

- Data are normally distributed

**How to read the test result:**

- It is based on P-value

| P-Value > 0.05 | There is no statistically significant difference between the two population means |
|---|---|
| P-Value ≤ 0.05 | The two population means are significantly different |

# Hypothesis Testing: 2-Sample t

MINITAB:

### Stat > Basic Statistic > 2-Sample t…

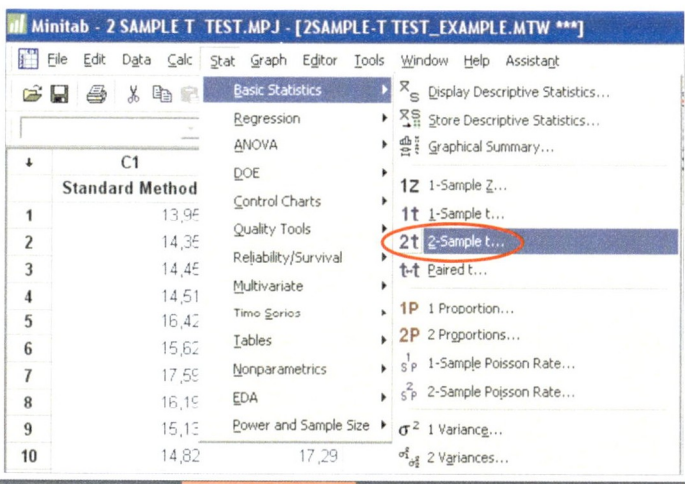

# Hypothesis Testing: 2-Sample t

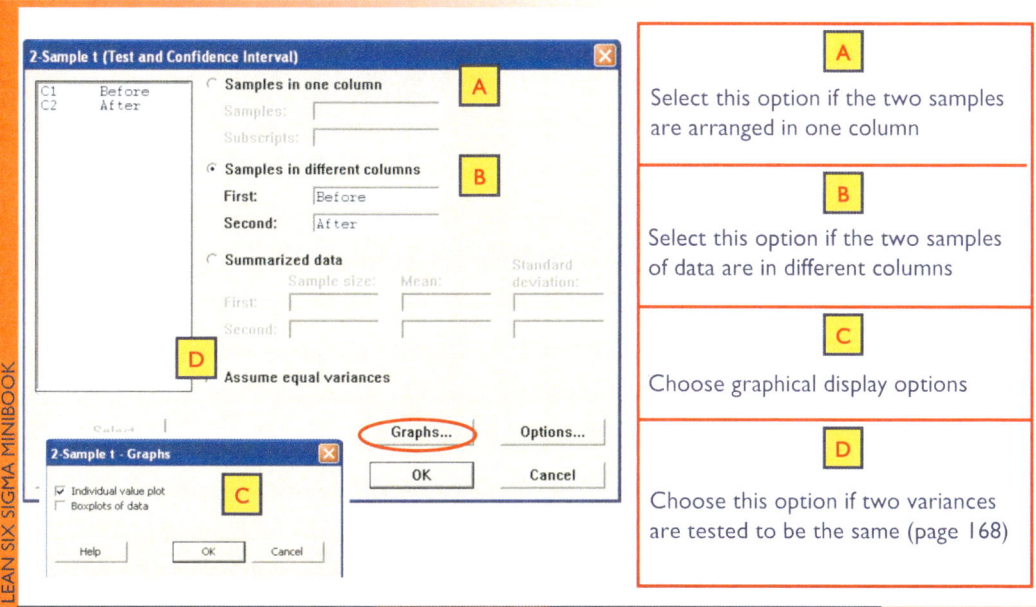

| | |
|---|---|
| **A** | Select this option if the two samples are arranged in one column |
| **B** | Select this option if the two samples of data are in different columns |
| **C** | Choose graphical display options |
| **D** | Choose this option if two variances are tested to be the same (page 168) |

# Hypothesis Testing: 2-Sample t

## MINITAB: Output

**Individual Value Plot of Before; After**

Two-Sample T-Test and CI: Before; After

Two-sample T for Before vs After

|  | N | Mean | StDev | SE Mean |
|---|---|---|---|---|
| Before | 75 | 75,21 | 1,66 | 0,19 |
| After | 75 | 77,08 | 1,62 | 0,19 |

Difference = mu (Before) - mu (After)
Estimate for difference: -1,874
95% CI for difference: (-2,403; -1,344)
T-Test of difference = 0 (vs not =): T-Value = -6,99   P-Value = 0,000  DF = 147

### A
It is visually possible to observe if two groups have equal means (two clusters of points will be at the same height if means are equal)

### B
If the P value is less than 0.05, it indicates that two population means are statistically and significantly different

# Hypothesis Testing: Paired t-Test

Objective:

- Paired t-Test is a hypothesis test that compares the **mean** differences of two related samples (**paired** samples). It is useful when we want to eliminate the presence of significant differences among sample members (person, machine, etc.) from the tests. For example, we can use Paired t-Test to compare the effects of two different machines (Machine A and B) on the same part

- The Paired t-Test would likely create tighter Confidence Intervals, because only the variation in paired difference is considered

Fundamental Assumptions:

- Two data sets have to be matched

- The distribution of the difference of data must be normal

# Hypothesis Testing: Paired t-Test

Example of *paired* data collection:

| C1 | C2 | C3 | C4 |
|---|---|---|---|
| Part | Measurement System1 | Measurement System2 | Difference in Measurement |
| 1 | 10,4010 | 10,5100 | -0,109041 |
| 2 | 10,6420 | 10,7230 | -0,081006 |
| 3 | 10,2032 | 10,3200 | -0,116827 |
| 4 | 11,2434 | 11,0900 | 0,153416 |
| 5 | 11,2198 | 11,3560 | -0,136243 |
| 6 | 9,7822 | 10,0500 | -0,267777 |
| 7 | 10,1403 | 10,1567 | -0,016438 |
| 8 | 10,1583 | 10,4500 | -0,291710 |
| 9 | 12,0665 | 11,9700 | 0,096516 |
| 10 | 10,5171 | 10,5200 | -0,002857 |

How to read the test results:

- The testing result is based on P-Value

| P-Value > 0.05 | There is no statistically significant difference for two matched data sets |
|---|---|
| P-Value ≤ 0.05 | There is a statistically significant difference in two matched data sets |

LEAN SIX SIGMA MINIBOOK

# Hypothesis Testing: Paired t-Test

MINITAB:

$\underline{S}$tat > $\underline{B}$asic Statistic > $\underline{P}$aired t...

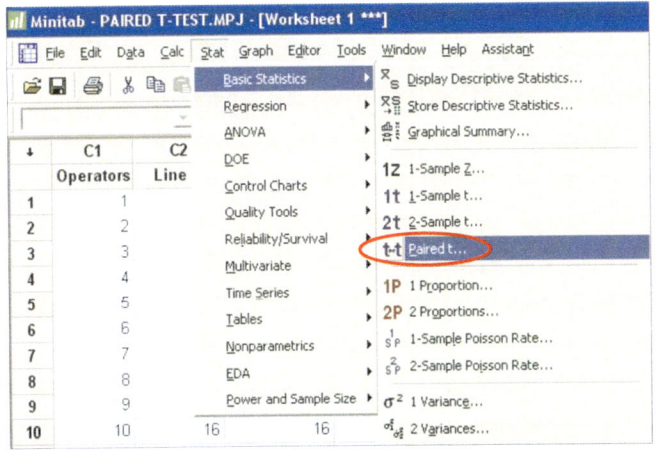

# Hypothesis Testing: Paired t-Test

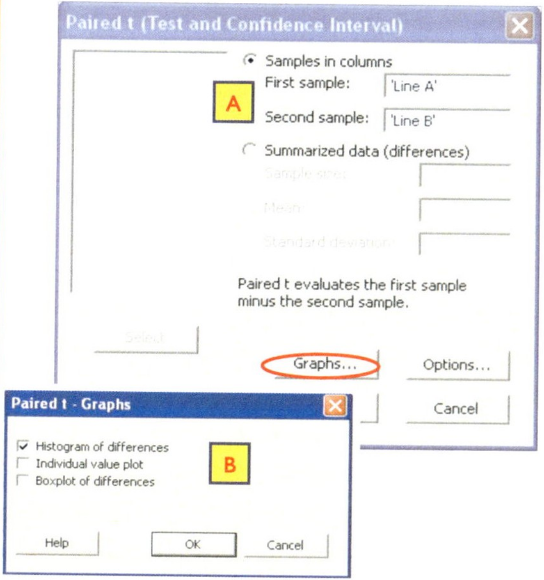

A

Select this option (*samples in columns*) if two data sets are in different columns and select columns for two samples

B

Select graphic display options for the test (e.g., select histogram of differences)

**Paired t (Test and Confidence Interval)**

- ⊙ Samples in columns
  - First sample: `'Line A'`
  - Second sample: `'Line B'`
- ○ Summarized data (differences)
  - Sample size:
  - Mean:
  - Standard deviation:

Paired t evaluates the first sample minus the second sample.

Graphs...    Options...

Cancel

**Paired t - Graphs**

- ☑ Histogram of differences
- ☐ Individual value plot
- ☐ Boxplot of differences

Help    OK    Cancel

# Hypothesis Testing: Paired t-Test

## MINITAB: Output

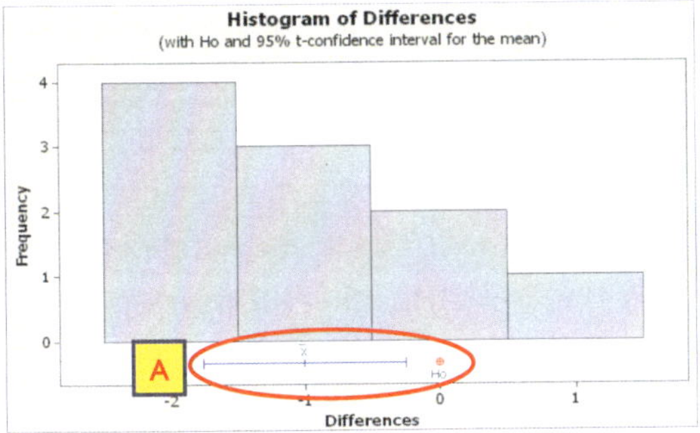

**Histogram of Differences**
(with Ho and 95% t-confidence interval for the mean)

Frequency

4

3

2

1

0

A

-2

-1

0

1

**Differences**

A

This graphic display is to check if the confidence interval contains zero. If yes, it indicates $H_0$ cannot be rejected i.e. there is no significant difference for two data pairs

B

If P-Value is $\leq 0.05$, it indicates that two data pairs are statistically different

**Paired T-Test and CI: Line A: Line B**

Paired T for Line A - Line B

|  | N | Mean | StDev | SE Mean |
|---|---|---|---|---|
| Line A | 10 | 12,300 | 3,093 | 0,978 |
| Line B | 10 | 13,300 | 2,908 | 0,920 |
| Difference | 10 | -1,000 | 1,054 | 0,333 |

95% CI for mean difference: (-1,754; -0,246)
T-Test of mean difference = 0 (vs not = 0): T-Value = -3,00  P-Value = 0,015

# Hypothesis Testing: ANOVA

## Objective:

- ANOVA is a Hypothesis Testing procedure that can compare the **means** of 2 or more groups
- ANOVA is also an analysis procedure that can perform key tests for other statistical methods such as Regressions and DOE

## Fundamental Assumptions:*

- Data is normally distributed
- Variances are the same for all groups (page 168)

| P-Value > 0.05 | There is no significant differences among group means |
|---|---|
| P-Value ≤ 0.05 | At least one group mean is significantly different than others |

* Residuals analysis can be used to identify the goodness of the analytical model (also used in Regression, page186)

# Hypothesis Testing: ANOVA

## MINITAB:

### <u>S</u>tat > <u>A</u>NOVA > <u>O</u>ne-Way...

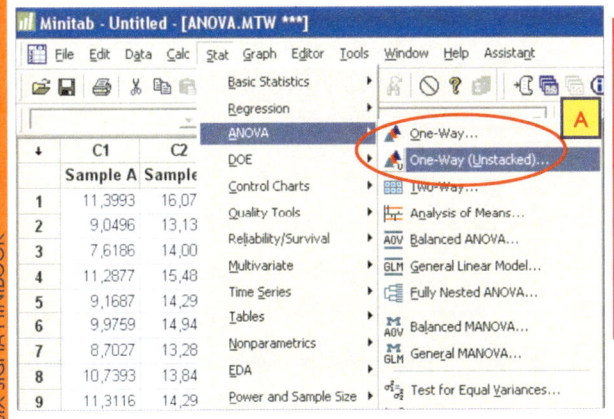

Select:

- *One Way (Unstacked)*. When sample data are placed in different columns

- *One-Way*. When all sample data are placed in one column

# Hypothesis Testing: ANOVA

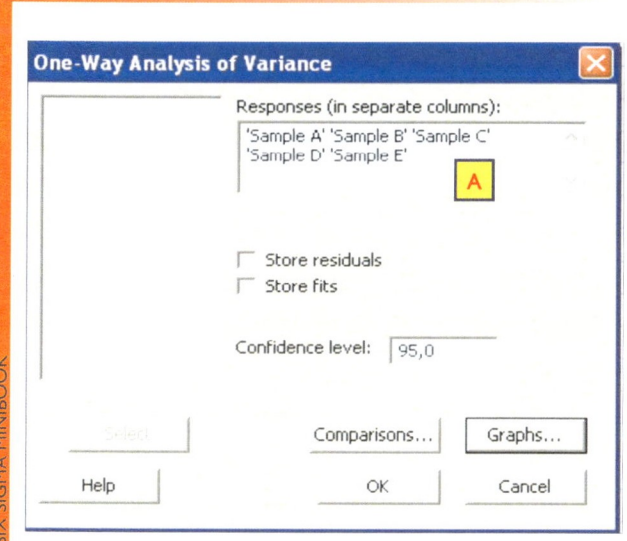

Enter the response columns to be analyzed

LEAN SIX SIGMA MINIBOOK

# Hypothesis Testing: ANOVA

## MINITAB: Output

One-way ANOVA: Sample A; Sample B; Sample C; Sample D; Sample E

| Source | DF | SS | MS | F | P |
|--------|-----|--------|--------|--------|-------|
| Factor | 4 | 598,52 | 149,63 | 110,80 | 0,000 |

| | | | | |
|--------|-----|--------|--------|
| Error | 100 | 135,05 | 1,35 |
| Total | 104 | 733,57 | |

S = 1,162   R-Sq = 81,59%   R-Sq(adj) = 80,85%

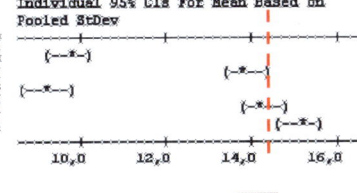

Individual 95% CIs For Mean Based on Pooled StDev

| Level | N | Mean | StDev |
|----------|-----|--------|-------|
| Sample A | 20 | 9,718 | 1,064 |
| Sample B | 20 | 13,821 | 1,410 |
| Sample C | 15 | 9,201 | 1,005 |
| Sample D | 25 | 14,285 | 1,218 |
| Sample E | 25 | 15,152 | 1,041 |

Pooled StDev = 1,162

**A**

When P-Value is ≤ 0.05, it indicates that at least one population mean is statistically and significantly different than others

**B**

Visually, overlapping Confidence Intervals (indicated by the red line crossing 3 groups) highlights there is no significant difference on means for these groups. Non-overlapping Confidence Intervals indicates significant differences

# Hypothesis testing: Chi-Square

Objective:

• The Chi-Square test aims to compare proportions or frequencies of occurrences by several groups

Note:

• Normal distribution of data is not required

| P-Value > 0.05 | There is no difference between the proportions (frequencies) of samples |
| P-Value $\leq$ 0.05 | The proportion (frequencies) of at least one group is statistically different than others |

# Hypothesis Testing: Chi-Square

MINITAB:

Stat > Tables > Chi-Square Test (Table in Worksheet)…

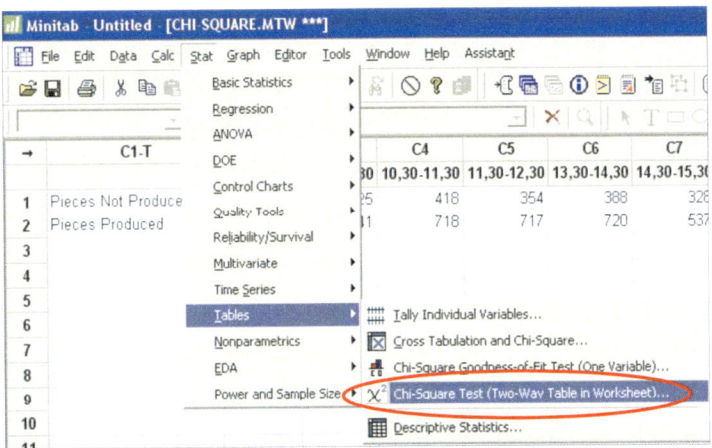

# Hypothesis Testing: Chi-Square

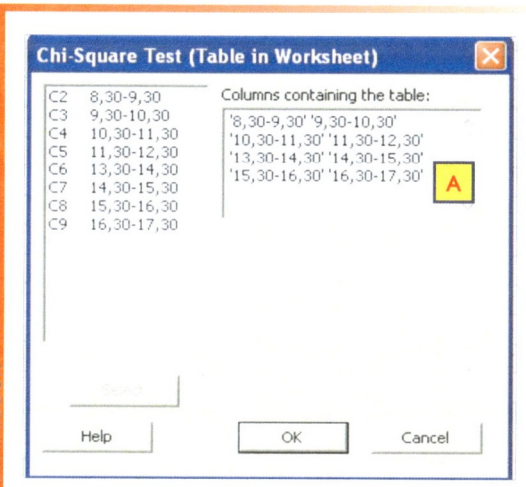

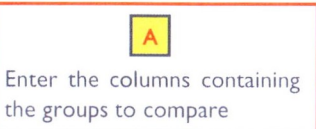

A

Enter the columns containing the groups to compare

# Hypothesis Testing: Chi-Square

## MINITAB: Output

Chi-Square Test: 8.30-9.30; 9.30-10.30; 10.30-11.30; 11.30-12.30; 13.30-14.30;

Expected counts are printed below observed counts
Chi-Square contributions are printed below expected counts

| | 8,30-9,30 | 9,30-10,30 | 10,30-11,30 | 11,30-12,30 | 13,30-14,30 |
|---|---|---|---|---|---|
| 1 | 378 | 325 | 418 | 354 | 388 |
| | 384,48 | 306,03 | 401,44 | 378,47 | 391,55 |
| | 0,109 | 1,176 | 0,683 | 1,582 | 0,032 |
| 2 | 710 | 541 | 718 | 717 | 720 |
| | 703,52 | 559,97 | 734,56 | 692,53 | 716,45 |
| | 0,060 | 0,643 | 0,373 | 0,865 | 0,018 |
| Total | 1088 | 866 | 1136 | 1071 | 1108 |

| | 14,30-15,30 | 15,30-16,30 | 16,30-17,30 | Total |
|---|---|---|---|---|
| 1 | 328 | 380 | 308 | 2879 |
| | 305,68 | 388,72 | 322,64 | |
| | 1,630 | 0,196 | 0,664 | |
| 2 | 537 | 720 | 605 | 5268 |
| | 559,32 | 711,28 | 590,36 | |
| | 0,891 | 0,107 | 0,363 | |
| Total | 865 |  1100 | 913 | 8147 |

Chi-Sq = 9,392; DF = 7; P-Value = 0,226

**A**

Highlighted Items:

- The calculated Chi Square statistic value by the formula:

$$\chi^2 = \sum \frac{(Observed - Expected)^2}{Expected}$$

- P-Value ≤ 0.05 is needed to indicate statistical significance between groups. For example, P-Value of 0.226 indicates there is no significant differences among groups

# Hypothesis Testing: Test for Equal Variances

Objective:

- This test compares Variances of several groups to determine if they are equal

Fundamental Assumptions:

- It is necessary to choose the most appropriate method based on the distribution of data:
  - Normally distributed data → *F-Test* (for 2 samples) and *Bartlett's Test* (for more than 2 samples)
  - General continuous distribution → *Levene's Test*

| P-Value > 0.05 | There is no statistically significant difference on Variances of groups |
| P-Value ≤ 0.05 | There is at least one group Variance that is significantly different from others |

# Hypothesis Testing: Test for Equal Variances

MINITAB:

    <u>S</u>tat > <u>B</u>asic Statistics > Test for Equal <u>V</u>ariances…

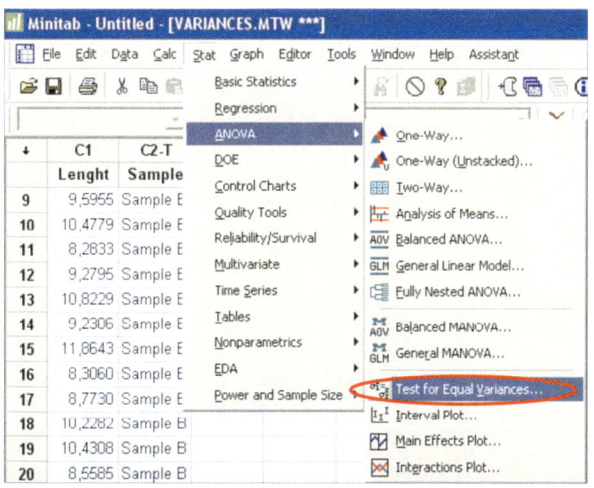

# Hypothesis Testing: Test for Equal Variances

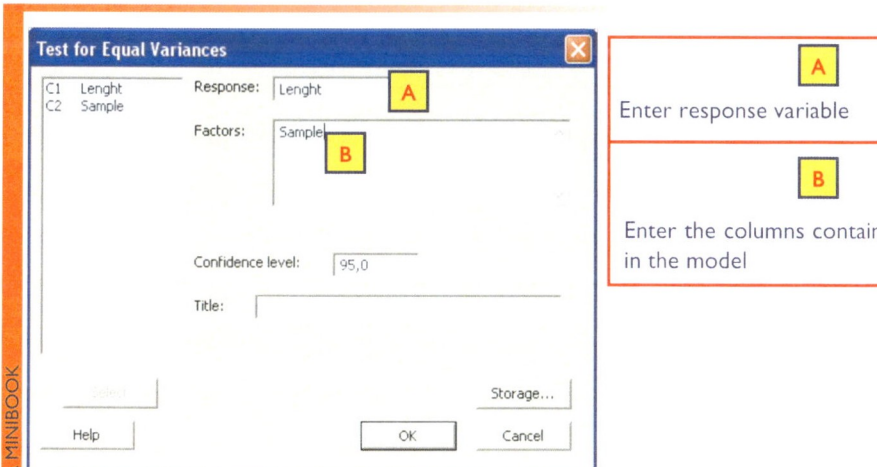

# Hypothesis Testing: Test for Equal Variances

## MINITAB: Output

**Test for Equal Variances: Length versus Sample**

95% Bonferroni confidence intervals for standard deviations

| Sample | N | Lower | StDev | Upper |
|--------|-----|---------|---------|---------|
| Sample A | 20 | 0,923465 | 1,31594 | 2,19260 |
| Sample B | 20 | 0,875280 | 1,24728 | 2,07820 |
| Sample C | 15 | 0,729982 | 1,09183 | 2,02382 |
| Sample D | 20 | 0,763471 | 1,08795 | 1,81273 |
| Sample E | 20 | 0,482262 | 0,68723 | 1,14505 |

Bartlett's Test (Normal Distribution)
Test statistic = 8,14; p-value = 0,087

Levene's Test (Any Continuous Distribution)
Test statistic = 1,79; p-value = 0,139

**Test for Equal Variances: Length versus Sample**

Minitab will choose F-Test or Bartlett's Test depending on the number of groups. We have to choose the right test (Bartlett or Levene) based on the distribution of data (normal or generic continuous distribution)

We need to see P-Value ≤ 0.05 in order to determine whether there is a significant difference in group variances

# Hypothesis Testing: Test for Equal Variances

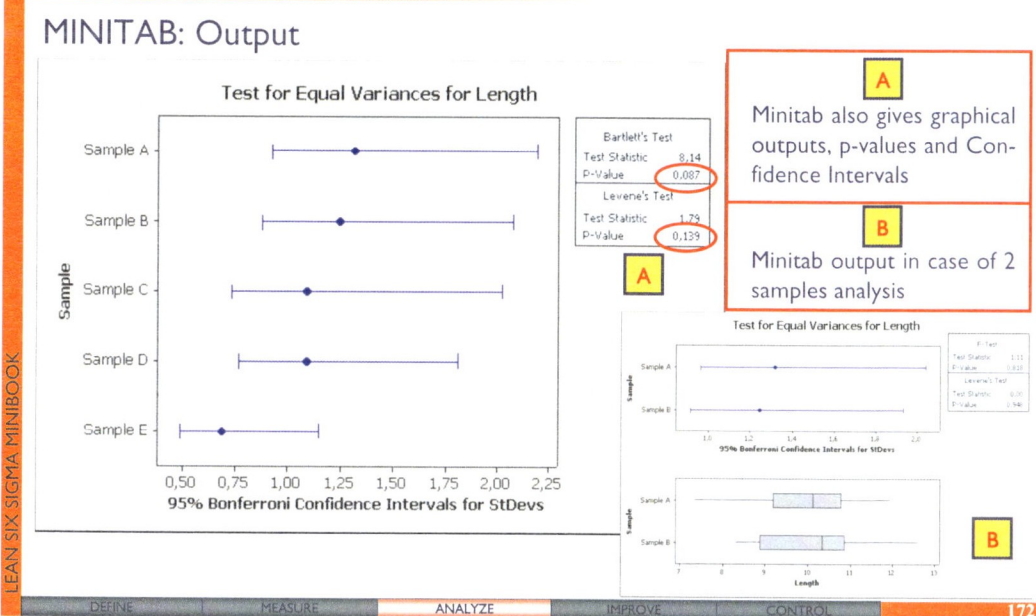

## MINITAB: Output

**A** Minitab also gives graphical outputs, p-values and Confidence Intervals

**B** Minitab output in case of 2 samples analysis

# Hypothesis Testing & Minitab

Minitab Assistant helps you to choose the right Hypothesis Testing:

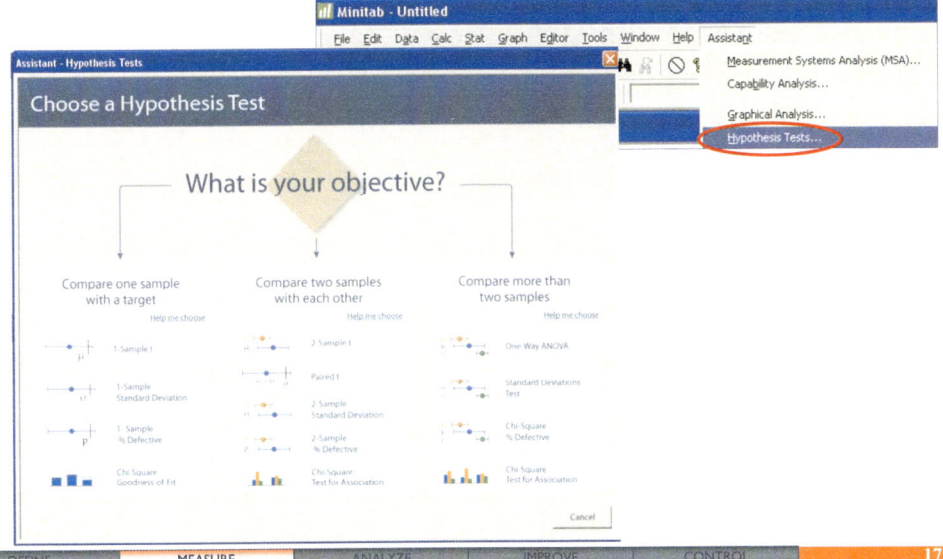

# Scatter Diagram

Objective:

- *Scatter Diagram* is a graph that can be used to determine a possible correlation between a pair of input and output variables
- There are 3 possible situations:

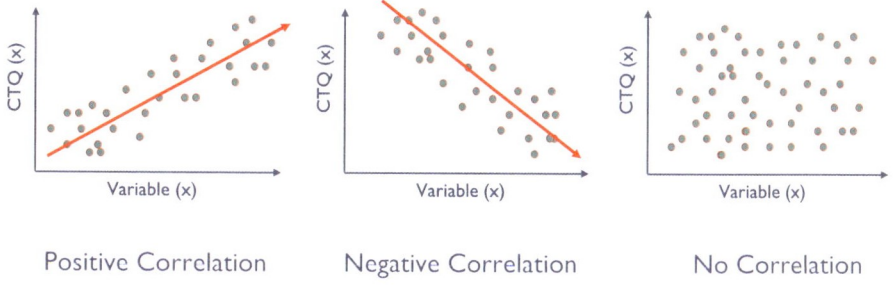

Positive Correlation      Negative Correlation      No Correlation

# Scatter Diagram

MINITAB:

## Graph> Scatterplot…

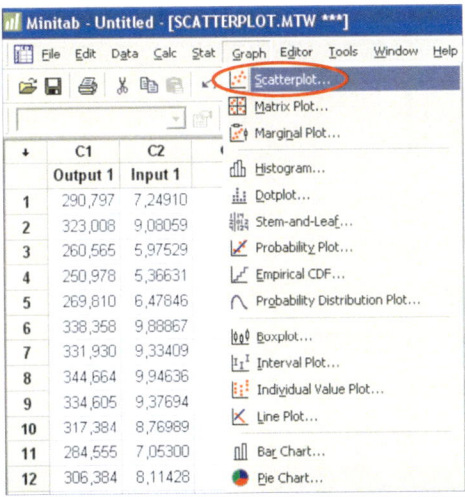

# Scatter Diagram

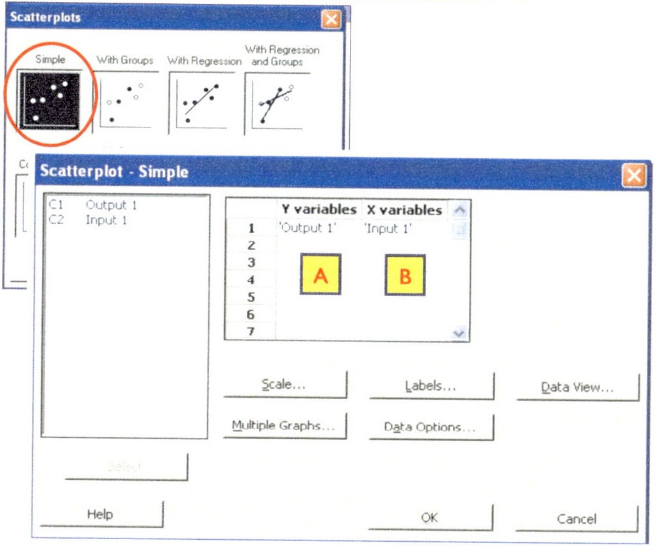

**A**

Enter the column that contains output variable (Response or Y *variable*)

**B**

Enter the column that contains input variable (*X variable*)

# Scatter Diagram

## MINITAB: Output

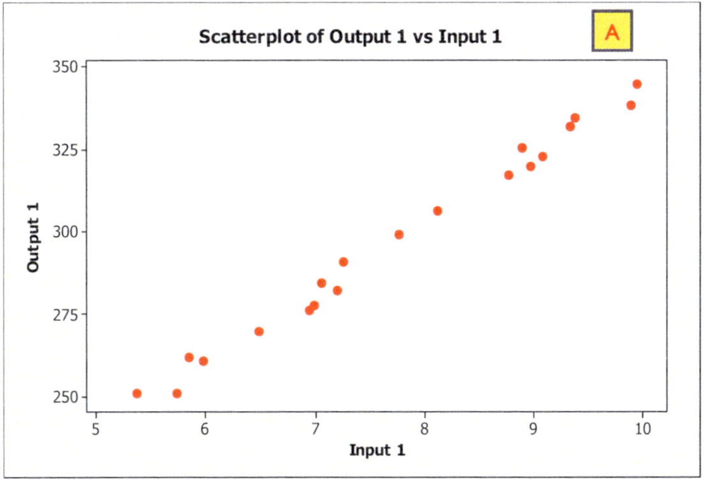

**Scatterplot of Output 1 vs Input 1**

*Scatter Diagram* is a graphic plot that displays Input-Output data pairs in X-Y axes

This plot can be used to qualitatively observe if there is a dependency relationship between 2 variables (potential correlation)

LEAN SIX SIGMA MINIBOOK

# Regression: Fitted Line Plot

### Objective:

* Regression is an analytical tool that can be used to establish, if it exists, a **mathematical model** between Input and Output variables (see Analytical Approach, page 183)

### Fundamental Assumptions:

* Variable Y $\rightarrow$ Continuous
* Variable X $\rightarrow$ Continuous
* Residuals $\rightarrow$ random variable

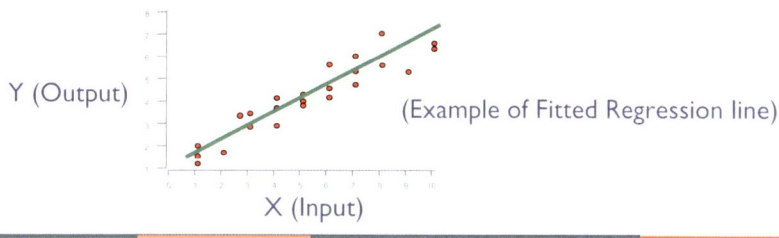

Y (Output)

(Example of Fitted Regression line)

X (Input)

# Regression: Fitted Line Plot

- For a Fitted Regression model, there is a commonly used performance indicator that can measure how good this model fits the data
- This indicator, called R-Squared (R-Sq%, or R-Sq), tells you the percentage of variations in data (from 1% to 100%) that can be explained by the regression model and is calculated by:

$$R\text{-}Sq = \frac{\text{explained variation}}{\text{total variation}}$$

| | |
|---|---|
| R-Sq ≥ 70 | The mathematical model explains the data correlation well |
| R-Sq < 70 | The mathematical model doesn't explain the data correlation well |

# Regression: Fitted Line Plot

MINITAB:

## $\underline{S}$tat > $\underline{R}$egression > $\underline{F}$itted Line Plot…

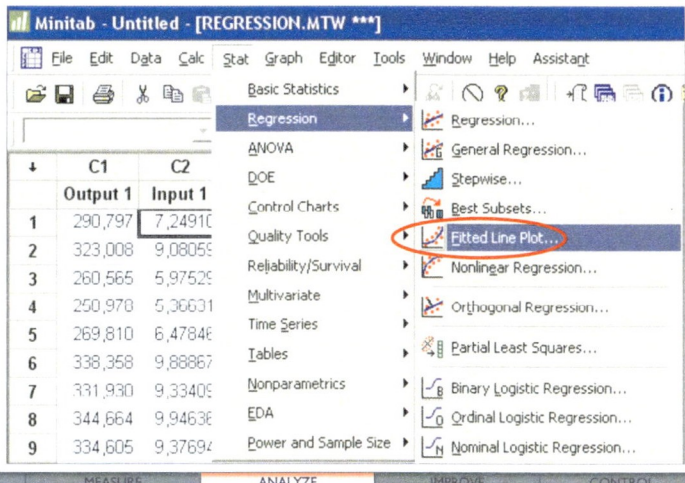

# Regression: Fitted Line Plot

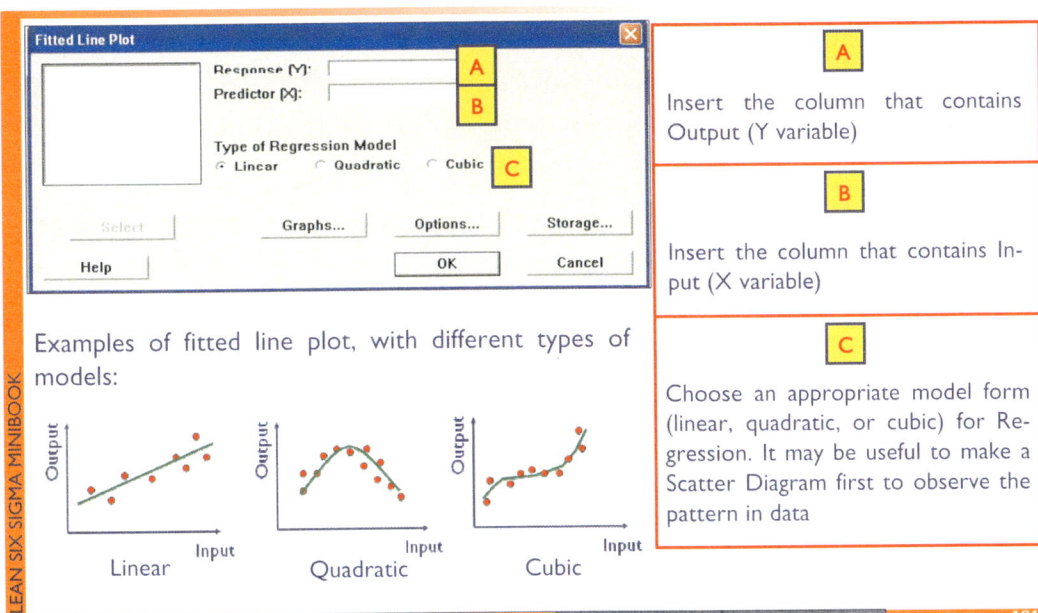

**Fitted Line Plot**

Response [Y]: ⬚ **A**
Predictor [X]: ⬚ **B**

**Type of Regression Model**
⦿ Linear    ○ Quadratic    ○ Cubic    **C**

Select    Graphs...    Options...    Storage...

Help    OK    Cancel

Examples of fitted line plot, with different types of models:

Linear    Quadratic    Cubic

**A**

Insert the column that contains Output (Y variable)

**B**

Insert the column that contains Input (X variable)

**C**

Choose an appropriate model form (linear, quadratic, or cubic) for Regression. It may be useful to make a Scatter Diagram first to observe the pattern in data

LEAN SIX SIGMA MINIBOOK

# Regression: Fitted Line Plot

## MINITAB: Output

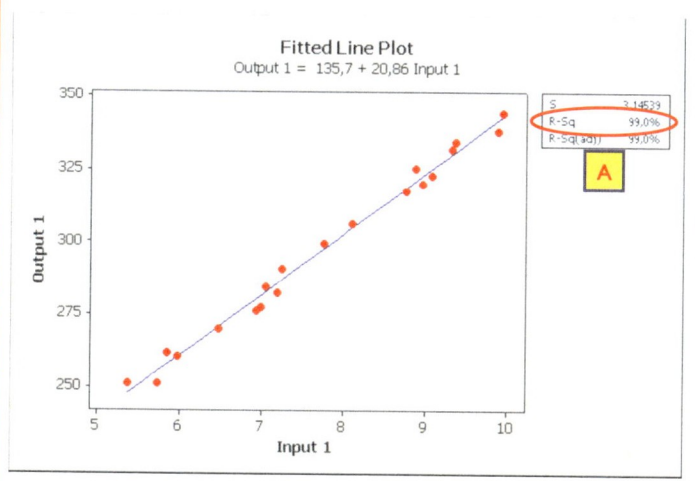

**Fitted Line Plot**
Output 1 = 135,7 + 20,86 Input 1

| S | 3.14539 |
| R-Sq | 99,0% |
| R-Sq(adj) | 99,0% |

**A**

Minitab provides the graphical illustration of a Fitted Regression line model, a Scatter Diagram of all data as well as calculated R-squared value

# Regression: Analytical Approach

MINITAB:

### Stat > Regression > Regression…

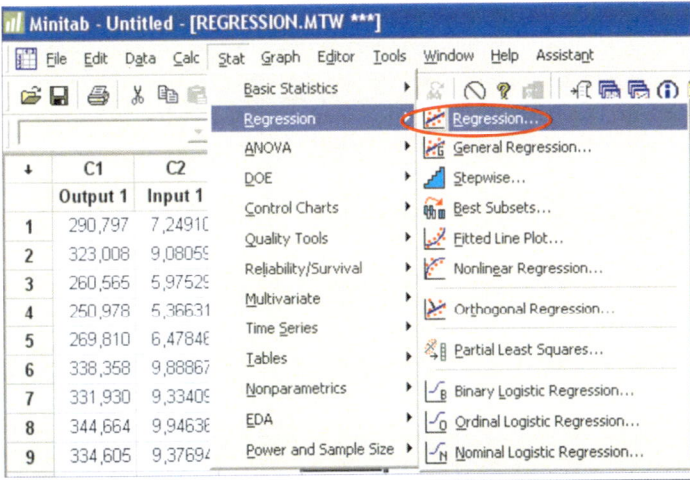

# Regression: Analytical Approach

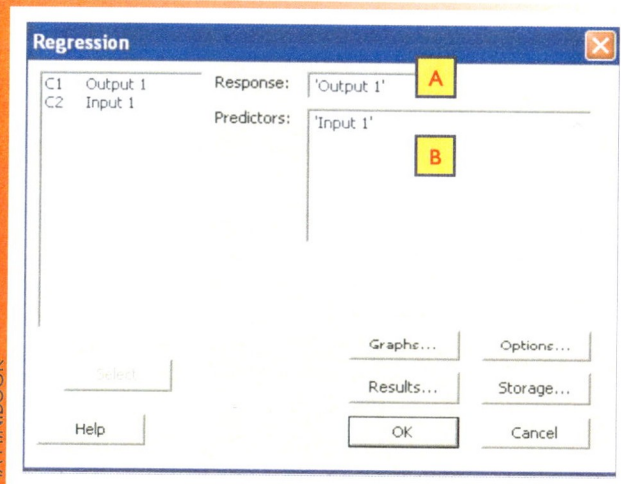

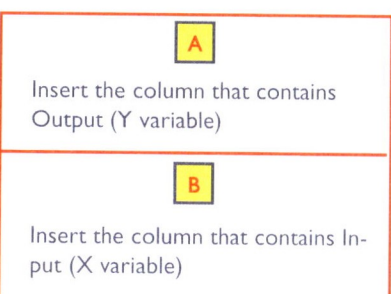

**A**

Insert the column that contains Output (Y variable)

**B**

Insert the column that contains Input (X variable)

LEAN SIX SIGMA MINIBOOK

# Regression: Analytical Approach

## MINITAB: Output

**Regression Analysis: Output 1 versus Input 1**

The regression equation is
Output 1 = 136 + 20,9 Input 1    **A**

| Predictor | Coef | SE Coef | T | P |
|---|---|---|---|---|
| Constant | 135,747 | 3,880 | 34,98 | 0,000 |
| Input 1 | 20,8559 | 0,4926 | 42,34 | 0,000 |

**B**

S = 3,14539    R-Sq = 99,0%    R-Sq(adj) = 99,0%

**C**

Analysis of Variance

| Source | DF | SS | MS | F | P |
|---|---|---|---|---|---|
| Regression | 1 | 17736 | 17736 | 1792,66 | 0,000 |
| Residual Error | 18 | 178 | 10 | | |
| Total | 19 | 17914 | | | |

**D**

---

**A**
Explicit Linear Regression equation with estimated coefficients

**B**
Significance tests of coefficients. If P-Value is ≤ 0.05, then the predictors are significant

**C**
R-sq value is calculated, and indicates how well the Regression model fits the data

**D**
In this test, if P-Value is ≤ 0.05 then it indicates that the Regression relationship is significant

# Regression: Assumptions

Residual analysis  (also applicable for Fitted Line Plot):
- Residuals not be related to X
- Residuals should be independent of time
- Residuals should be near constant regardless of predicted  Y values
- Residuals should be normally distributed

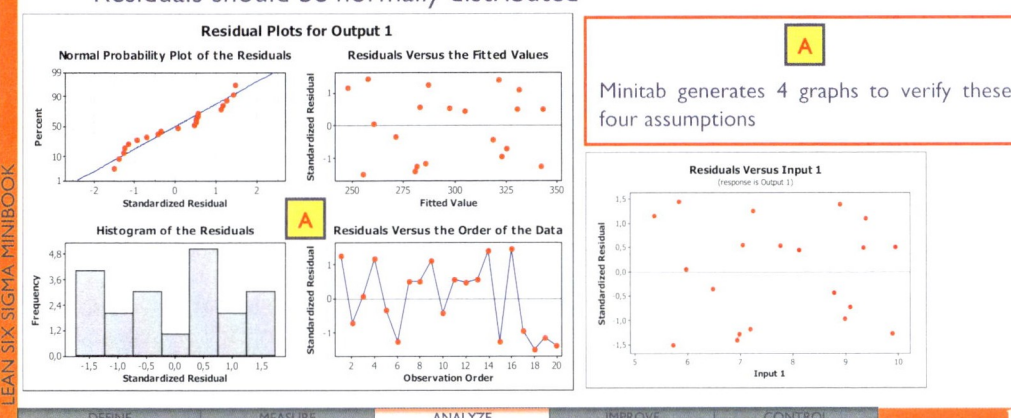

LEAN SIX SIGMA MINIBOOK

# IMPROVE

Arcidiacono G., Calabrese C., Yang K.: Leading processes to lead companies: Lean Six Sigma.
DOI 10-1007/978-88-470-2492-2, © Springer-Verlag Italia 2012

# IMPROVE

The *Improve* phase is the fourth step of the DMAIC Lean Six Sigma roadmap. In this step the existing process will be changed and optimized:

- This process optimization will be based on sound data analysis, a thorough understanding of the relationship between key process responses/performance metrics and key process variables, so this process optimization will be more likely to achieve real results backed by statistical confidence

- The improvement will take into account Lean applications, typical of Lean methodology and mindset, in order to reduce waste and to increase process efficiency

- The process optimization is based on scientific approach, real and accurate data. It is not based on subjective judgments

# 5S Program

## Objective:

- The "5S Program" is a system for creating and maintaining a work environment clean, orderly, efficient and safe. The benefits of this method can be evaluated in terms of Quality, Safety and Productivity:

| Productivity | Safety | Quality |
|---|---|---|
| • Eliminate wastes of time looking for equipment or items necessary for the job<br>• Reduce cycle times<br>• Maintain efficient equipment through proper maintenance and cleaning | • Reduce the likelihood of accident<br>• Making the workplace more ergonomic and comfortable | • Eliminate the possibility of using parts previously discarded<br>• Eliminate the possibility of using inappropriate tools |

## Overview:

# 5S Program

| STEP 5S | OBJECTIVE | PRACTICAL NOTES |
|---------|-----------|-----------------|
| SORT | Identify what is needed and what is not needed in the workplace. Eliminate or segregate what is not necessary | Try to answer these questions: What is the use of this object? Why do I have it? How often is it used? |
| SET IN ORDER | Organize and arrange everything you need in the workplace so it can be quickly found, used and stored | Place all necessary items in the best possible location, at the "point of use" Use the 'visual' area approach (standard and not standard must be easily identified) Use labels and boards to make clear the inventory, equipment and other items so that everybody can understand the workplace |
| SHINE | Clean and maintain order in the workplace, equipment, floors into the shopfloor/office | Clean inside, under, over and around machinery and furniture. The cleaning of machine is a very important point because it can help prevent damage before it happens |
| STANDARDISE | Maintain and improve the standard of the first 3S | Introduce changes to the workstations that make cleaning and removal of dirt at the root easy and quick Use checklist to perform daily tasks of cleaning, maintenance and organization Identify areas and responsibilities |
| SUSTAIN | Make the standard 5S a daily habit and part of everyday work | Strengthen the workstation habits according to the 5S approach. Use audits with a "Steering Team" and relative corrective actions. Continuous improvement of 5S Use, if necessary, the OPL (One Point Lesson) and information board |

# 5S Program

How to conduct a 5S event in practical steps:

1. Identify the area where the 5S system (production, warehouse, shipping, quality, laboratory) is implemented
2. Divide the area into zones and related people in charge
3. Define a *Steering Team*\*
4. Define roles of area responsible
5. Identify Steering Team responsibilities
6. Define the 5S Checklist and audit standard format
7. Establish the terms and timetable for implementing audit
8. Implement the 5S information board

\* See the examples of format in Kaizen standard form session (page 310)

# 5S Program

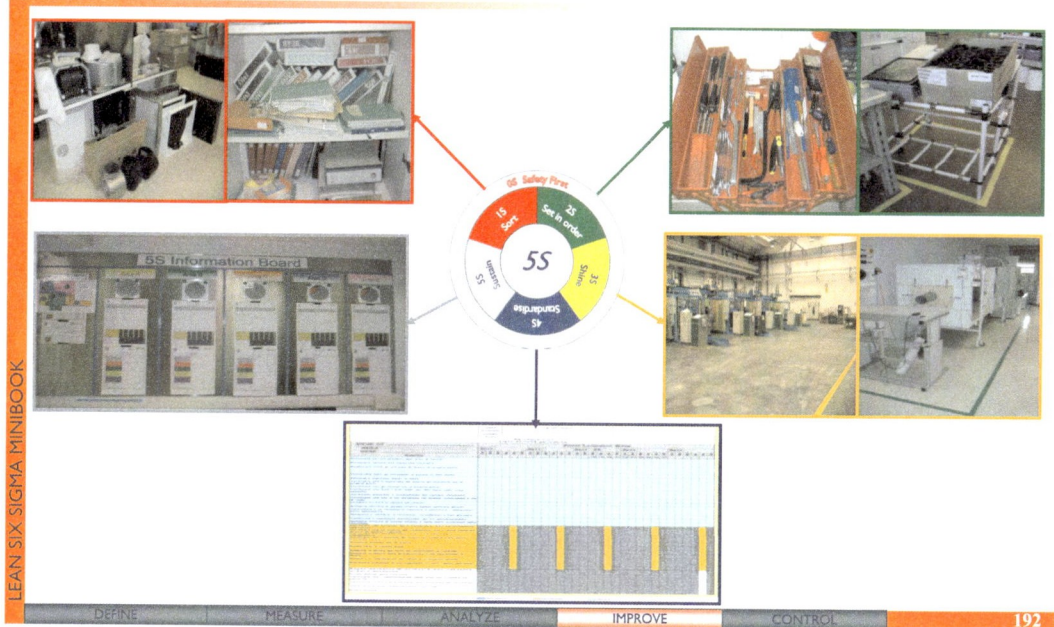

# 5S Program

## How to measure 5S performance:

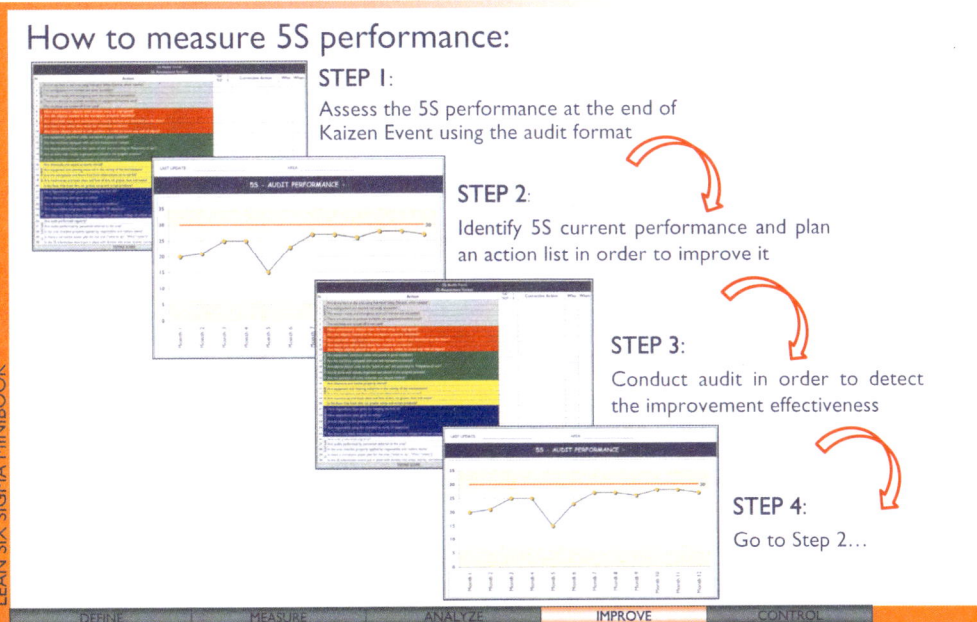

**STEP 1**:

Assess the 5S performance at the end of Kaizen Event using the audit format

**STEP 2**:

Identify 5S current performance and plan an action list in order to improve it

**STEP 3**:

Conduct audit in order to detect the improvement effectiveness

**STEP 4**:

Go to Step 2…

# 5S Program

Some tips before getting started:

- For the success of 5S implementation go through employee involvement
- Make sure that employees understand the 5S system and why it is so important for them and for the company
- Start with a pilot phase and then extend the approach to other areas
- It is necessary in the first phase of implementation for everyone to sacrifice and appreciate the real benefits of the approach
- The manager should sponsor the activity
- Make responsibilities clear and understood by everybody
- Make the process as visual as possible
- Link the 5S program to all other Kaizen activities (e.g. Standard Work, SMED, TPM, Six Sigma projects etc.)
- If possible, integrate the 5S program with safety program

# Standard Work

## Objective:

- Standard Work [G] is the most effective combination of manpower, materials and machinery to produce something in the time, quality and quantity required by the customer. This is done through continuous observation and improvement of the workplace. Standard Work is characterized by three main elements: Takt Time; Standard Work in process and Work sequence

## Definition:

- **Standard Work in process**: The minimum work-in-process needed to maintain Standard Work. Standard WIP can be parts completed and in the machine after auto cycle, parts placed in equipment with cycle times bigger than Takt Time, and parts handled by the operators on the production line

- **Work sequence**: The sequence of steps and activities that need to be performed in order to complete the production process

- **Takt Time** (see page 120)

# Standard Work

## Step 1: analyze the operator Cycle Time

| Observation Date | 21/09/2010 | | Kaizen Leader | Mr. Green |
|---|---|---|---|---|
| Company | XFS Factory | | Kaizen Team | Mr. Yellow; Mrs.Pink; Mr Grey; Ms. Orange |
| Kaizen | Standard Work | | | |
| Observation Time | 10.50 AM | | | |
| Process | Assembly Line 23 | | | |

| | | **Cycle Time Observation Form** | | | | | | | | | | | | |
|---|---|---|---|---|---|---|---|---|---|---|---|---|---|---|
| N. | Task | Cycle 1 | Cycle 2 | Cycle 3 | Cycle 4 | Cycle 5 | Cycle 6 | Cycle 7 | Cycle 8 | Cycle 9 | Cycle 10 | Mean | Lowest repeatable | Standard time |
| 1 | Activity A | 8 / 8 | 7 / 7 | 10 / 10 | 11 / 11 | 9 / 9 | 9 / 9 | 10 / 10 | 11 / 11 | 8 / 8 | 8 / 8 | 9,1 | 8 | 9 |
| 2 | Activity B | 12 / 20 | 12 / 19 | 8 / 18 | 9 / 20 | 11 / 20 | 9 / 18 | 10 / 20 | 10 / 21 | 9 / 17 | 10 / 18 | 10 | 10 | 10 |
| 3 | Activity C | 6 / 26 | 9 / 28 | 6 / 24 | 6 / 26 | 8 / 28 | 9 / 27 | 10 / 30 | 8 / 29 | 9 / 26 | 9 / 27 | 8 | 6 | 8 |
| 4 | Activity D | 4 / 30 | 7 / 35 | 5 / 29 | 6 / 32 | 7 / 35 | 8 / 35 | 6 / 36 | 6 / 35 | 5 / 31 | 5 / 32 | 5,9 | 5 | 6 |
| 5 | Activity E | 6 / 36 | 6 / 41 | 5 / 34 | 7 / 39 | 6 / 41 | 7 / 42 | 6 / 42 | 5 / 40 | 6 / 37 | 6 / 38 | 6 | 6 | 6 |
| 6 | Activity F | 5 / 41 | 8 / 49 | 7 / 41 | 8 / 47 | 7 / 48 | 8 / 50 | 7 / 49 | 6 / 46 | 7 / 44 | 8 / 46 | 7,1 | 7 | 7 |
| 7 | Activity G | 3 / 44 | 5 / 54 | 4 / 45 | 6 / 53 | 7 / 55 | 5 / 55 | 4 / 53 | 7 / 53 | 4 / 48 | 5 / 51 | 5 | 4 | 5 |
| 8 | Activity H | 6 / 50 | 7 / 61 | 7 / 52 | 7 / 60 | 7 / 62 | 8 / 63 | 8 / 61 | 6 / 59 | 6 / 54 | 8 / 59 | 7 | 6 | 7 |
| 9 | Activity I | 10 / 60 | 12 / 73 | 12 / 64 | 13 / 73 | 10 / 72 | 10 / 73 | 11 / 72 | 10 / 69 | 12 / 66 | 11 / 70 | 11,1 | 10 | 11 |
| | Cycle time (1 Cycle) | 60 | 73 | 64 | 73 | 72 | 73 | 72 | 69 | 66 | 70 | | | 69 |

# Standard Work

## Step 1: analyze the operator Cycle Time

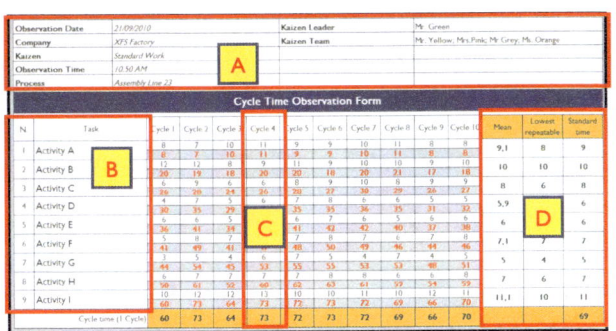

| Observation Date | 21/09/2010 | | Kaizen Leader | Mr. Green |
|---|---|---|---|---|
| Company | XFS Factory | | Kaizen Team | Mr. Yellow, Mrs Pink, Mr Grey, Ms. Orange |
| Kaizen | Standard Work | | | |
| Observation Time | 10.50 AM | | | |
| Process | Assembly Line 23 | | | |

**Cycle Time Observation Form**

| N | Task | Mean | Lowest repeatable | Standard time |
|---|---|---|---|---|
| 1 | Activity A | 9,1 | 8 | 9 |
| 2 | Activity B | 10 | 10 | 10 |
| 3 | Activity C | 8 | 6 | 8 |
| 4 | Activity D | 5,9 | | 6 |
| 5 | Activity E | 6 | | 6 |
| 6 | Activity F | 7,1 | 7 | 7 |
| 7 | Activity G | 5 | 4 | 5 |
| 8 | Activity H | 7 | 6 | 7 |
| 9 | Activity I | 11,1 | 10 | 11 |
| | Cycle time (1 Cycle) | | | 69 |

- Each task must be between 5 and 10 s
- Document the complete sequence of the operator
- Generally, in order to have a robust time, it is necessary to repeat the measurement from 7 to 10 times
- Use the chronometer and teach the operators how to measure by themselves
- During measurement try to identify potential improvements

Define the activity and the team involved in it

Identify the main tasks to complete a cycle

Measure, where possible, one entire cycle recording the single task time and the cumulative time

For each task identify mean and lowest repeatable performance. The time will be equal to the mean if the lowest repeatable is smaller than mean

LEAN SIX SIGMA MINIBOOK

DEFINE | MEASURE | ANALYZE | IMPROVE | CONTROL | 197

# Standard Work

## Step 2: understand the relation between Cycle Time and Takt Time

$$\text{TAKT TIME} = \frac{\text{Available Time (Time)}}{\text{Customer Demand (pcs/Time)}}$$

$$\text{N Operators necessary} = \frac{\text{Cycle Time}}{\text{Takt Time}} \quad \boxed{A}$$

**Available Time**: Total available time minus planned downtime (example: Breaks)

**Customer Demand**: Total expected demand from Customer (pieces per unit of time)

**Cycle Time**: Total manual working time for one cycle (Touch Time) plus automatic run time. In case of production mix, the weighted mean (weighted on quantity) can be useful

**CAUTION (LEAN METHODOLOGY BASIC CONCEPT):**

Customer demand increase → Takt Time Decrease (increment of speed rate) → No. Operator needed increase

Customer demand decrease → Takt Time Increase (reduction of speed rate) → No. Operator needed decrease

LEAN SIX SIGMA MINIBOOK

# Standard Work

## Step 3: measure machine process capacity

| Observation Date | 21.09.2010 | | | | | Opearting time per shift | | 27600 s |
|---|---|---|---|---|---|---|---|---|
| Company | XFS Factory | | | | | Shift No | | 1 |
| Part No | Cylinder | | | | | Daily demand | | 400 pieces |
| Process | Assembly Line 23 | | | | | Supervisor | | Mrs. Brown |

**Process Capacity Form**

| Step # | Process Description | Machine | Base Time | | Tool Change | | | Total Time (s) | Processing Capacity | Remarks |
|---|---|---|---|---|---|---|---|---|---|---|
| | | | Manual Time (s) | Machine Time (s) | # Pieces /Change | Change Time | Tool Change Time (s) | | | |
| 1 | Press | HL 345 | 3 | 15 | 100 | 180 | 1,8 | 19,8 | 1394 | |
| 2 | Drilling | D10 | 6 | 65 | 70 | 200 | 2,9 | 73,9 | 374 | Maximum capacity 374 pieces |
| 3 | Testing | TEST3 | 5 | 45 | 150 | 300 | 2.0 | 52,0 | 531 | |
| | | | | | | | | | | |
| | | | | | | | | | | |
| | | | | | | | | | | |
| | | Total: | 14 | | | | Maximum daily production: | | 374 | |

# Standard Work

## Step 3: measure machine process capacity

| Observation Date | 21.09.2010 | | | Operating time per shift | | 27600 s |
|---|---|---|---|---|---|---|
| Company | XFS Factory | | | Shift No | | 1 |
| Part No | Cylinder | | | Daily demand | | 400 pieces |
| Process | Assembly Line 23 | | | Supervisor | | Mrs. Brown |

**Process Capacity Form**

| Step # | Process Description | Machine | Base Time | | | Tool Change | | Total Time (s) | Processing Capacity | Remarks |
|---|---|---|---|---|---|---|---|---|---|---|
| | | | Manual Time (s) | Machine Time (s) | # Pieces /Change | Change Time | Tool Change Time (s) | | | |
| 1 | Press | HL 345 | 3 | 15 | 100 | 180 | 1.8 | 19.8 | 1394 | |
| | Drilling | D10 | 6 | 65 | 70 | 200 | 2.9 | 73.9 | 374 | Maximum capacity 374 pieces |
| | Testing | TEST3 | 5 | 45 | 150 | | 2.0 | 52.0 | 531 | |
| | | | | | | | | | | |
| | | | | | | | | | | |
| | | | | | | | | | | |
| | | Total | 14 | | | Maximum daily production | | | 374 | |

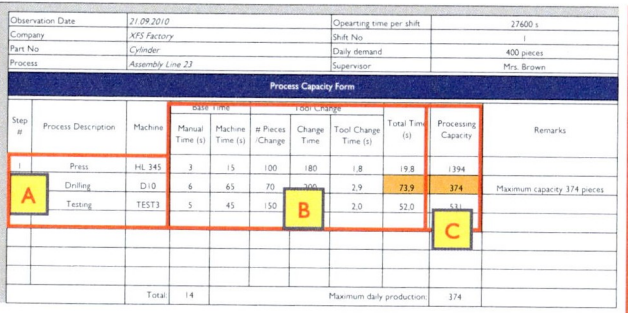

Identify machine process during the cycle

For each cycle, identify machine cycle time. Manual contribution and pieces change time to calculate one process cycle time

Calculate machine capacity as the ratio of operating time per shift and total time for each machine; example:
· Press Total Cycle time = 19.8 s
· Operating time = 27600 s
· Maximum capacity = 1394 pieces

- The capacity sheet is used to highlight machine capacity
- Identifies process bottlenecks: in the example Drilling Process has a maximum capacity of 374 pieces. At a daily demand of 400 pieces the machine is not capable of reaching customer needs
- Use one worksheet for each cell

# Standard Work

## Step 4: analyze interaction between operator and equipment

| Observation Date | 21.09.2010 | Opearting time per shift | 27600 s | - - - = Machine |
| Company | XFS Factory | Daily demand | 400 pieces | = Manual |
| Part No | Cylinder | Takt Time (s) | 69 | = Walking |
| Process | Assembly Line 23 | Supervisor | Mrs. Brown | = Takt Time |

### Standard Work Combination Form

| Step # | Task/Activity Description | Man. | Auto | Walk | Operation Working Time (s) |
|---|---|---|---|---|---|
| 1 | Activity 1 | | | 3 | |
| 2 | Activity 2 | 5 | | 2 | |
| 3 | Activity 3 | 2 | 60 | 3 | |
| 4 | Activity 4 | 15 | | 3 | |
| 5 | Activity 5 | 10 | | 2 | |
| 6 | Activity 6 | 2 | 30 | 2 | |
| 7 | Activity 7 | 20 | | | |

# Standard Work

## Step 4: analyze interaction between operator and equipment

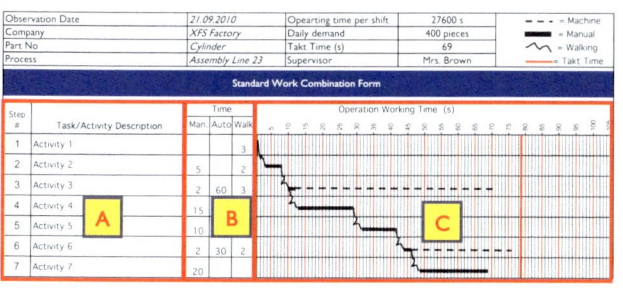

| Observation Date | 21.09.2010 | Opearting time per shift | 27600 s | - - - = Machine |
| Company | XFS Factory | Daily demand | 400 pieces | = Manual |
| Part No | Cylinder | Takt Time (s) | 69 | = Walking |
| Process | Assembly Line 23 | Supervisor | Mrs. Brown | = Takt Time |

**Standard Work Combination Form**

| Step # | Task/Activity Description | Time | | | Operation Working Time (s) |
|---|---|---|---|---|---|
| | | Man | Auto | Walk | |
| 1 | Activity 1 | | | 3 | |
| 2 | Activity 2 | 5 | | 2 | |
| 3 | Activity 3 | 2 | 60 | 3 | |
| 4 | Activity 4 | 15 | | | |
| 5 | Activity 5 | 10 | | | |
| 6 | Activity 6 | 2 | 30 | 2 | |
| 7 | Activity 7 | 20 | | | |

Identify the main steps (task/activity) necessary to complete one cycle

Identify for each step the operator, machine and walk time contribution

Identify on the sheet the operator steps, machine phases and walk steps. Highlight the Takt Time line

- The Standard Work combination sheet could be a good starting point for improvement identification
- Combines Man and equipment interactions
- Identifies the relation between Takt Time and process Cycle Time
- It could be useful to have one sheet for each operator

# Standard Work

## Step 5: map operator workload and make comparison with Takt Time

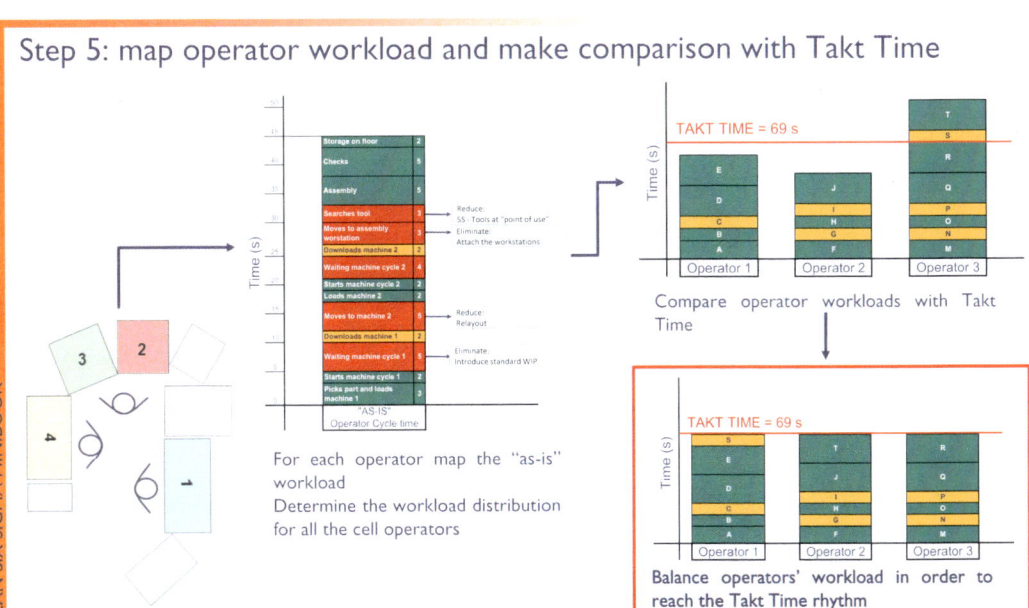

For each operator map the "as-is" workload

Determine the workload distribution for all the cell operators

Compare operator workloads with Takt Time

Balance operators' workload in order to reach the Takt Time rhythm

# Standard Work

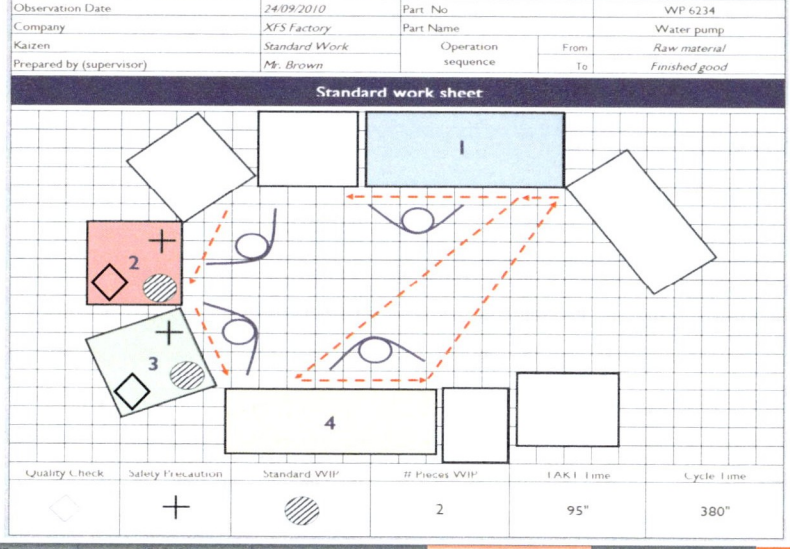

| Observation Date | 24/09/2010 | Part No | | WP 6234 |
|---|---|---|---|---|
| Company | XFS Factory | Part Name | | Water pump |
| Kaizen | Standard Work | Operation | From | Raw material |
| Prepared by (supervisor) | Mr. Brown | sequence | To | Finished good |

**Standard work sheet**

| Quality Check | Safety Precaution | Standard WIP | # Pieces WIP | TAKT Time | Cycle Time |
|---|---|---|---|---|---|
| ◇ | + | ▨ | 2 | 95" | 380" |

LEAN SIX SIGMA MINIBOOK

# Standard Work

## Step 6: prepare Standard Work sequence configuration according to Takt Time

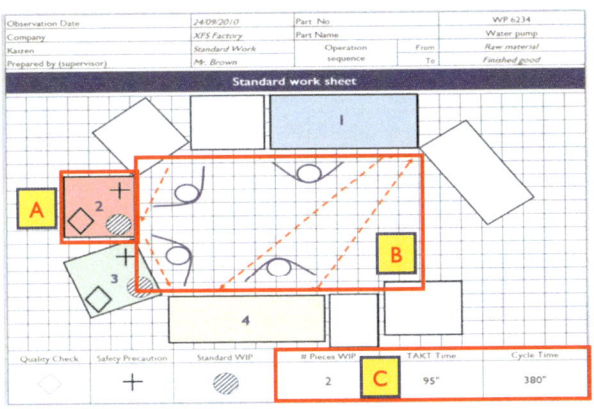

| Observation Date | 24/08/2010 | Part No | | WP 6234 |
| Company | X75 Factory | Part Name | | Water pump |
| Kaizen | Standard Work | Operation | From | Raw material |
| Prepared by (supervisor) | Mr. Brown | sequence | To | Finished good |

**Standard work sheet**

| Quality Check | Safety Precaution | Standard WIP | # Pieces WIP | TAKT Time | Cycle Time |
| ◇ | + | ▨ | 2 | 95" | 380" |

· It could be useful to have one sheet for each configuration of the cell
· Involve the team during sheet preparation

For each workstation identify Standard Work-in-process (the minimum work-in-process needed to maintain Standard Work), safety precautions and quality control

In this part of the sheet, highlight the movements of the operators involved in the process

The configuration identifies:
- No. Pieces WIP (in the example = 2)
- Takt Time (in the example = 95")
- Cycle time (in the example = 380")

# Standard Work

## Step 7: balance workload according to Customer Demand

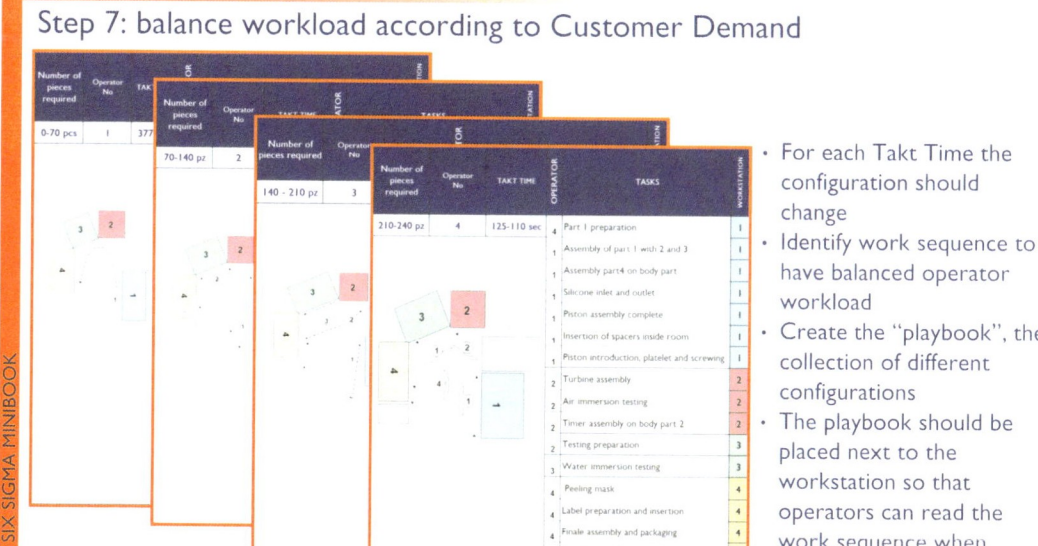

- For each Takt Time the configuration should change
- Identify work sequence to have balanced operator workload
- Create the "playbook", the collection of different configurations
- The playbook should be placed next to the workstation so that operators can read the work sequence when necessary

LEAN SIX SIGMA MINIBOOK

DEFINE · MEASURE · ANALYZE · IMPROVE · CONTROL

# Standard Work

## Japanese experience:

- At the beginning of the 20th century, especially in the USA, industrial engineers designed and implemented the standards of work, which often caused tense labor-management relationships. After the 1950s, in Japan, work standards were run by "quality circles" and constantly improved and revised by workers. This practice was proven to be very effective and empowering (Imai, 1997)

## Features of a good Standard Work (Imai, 1997):

1. Standards are the best, easiest, and safest way to do a job
2. They preserve know-how and expertise. Years of experience and knowledge can be lost by the loss of employees
3. They provide a way to measure performance
4. Correct standards show the relationship between cause and effect, leading to desired effects
5. Standards provide a basis for maintenance and improvement
6. They provide a set of visual signs on how to do the job
7. Standards are a basis for training
8. They are a basis for auditing
9. They are a mean to prevent recurrence of errors
10. Standards minimize variability

# Cell Design

## What is a "Cell":

- A "Cell$^{(G)}$" is a workplace in which equipment, people, machinery, materials and methods are arranged to have a continuous production flow (Continuous Flow). It allows the "One piece flow" principle: an operator can process the entire product from beginning to end without interruption. The cell generally runs for a family of products. A typical configuration is the "U" shape.

## Overview:

From traditional line to
CELL DESIGN

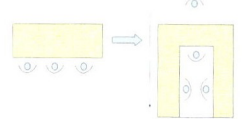

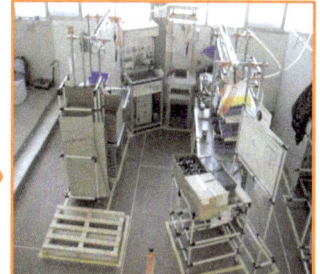

Full-size cardboard cutouts

Cell Layout example

# Cell Design

## U shape cell advantages:

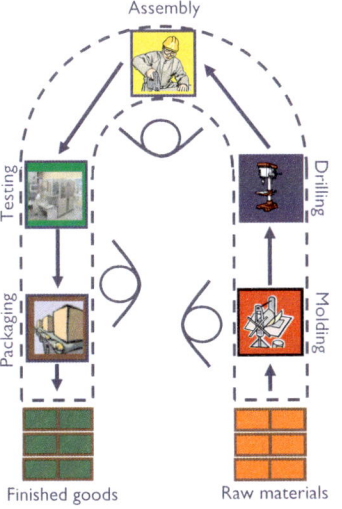

- Allows a better balance of workload
- Promotes Visual Management approach
- Reduces movements along the line
- Improves communication
- Allows multi-processing
- Identifies abnormalities easily
- Does not interrupt the production inside the cell through the supply of materials from outside, thanks to "water spider"
- More flexibility in changing the Takt rhythm

# Cell Design

## How to implement a "cell"? Follow the operating steps below:

- Categorize products in "Product Family"
- Determine Takt Time for each family (using, if necessary weighted mean)
- Make a study of the elementary steps and relative times
- Workload balance and Standard Work
- Design the "U" shape cell in terms of:
  - Layout & workstation
  - Number of operators necessary to reach customer demand
  - Movements
  - Materials management and Standard WIP

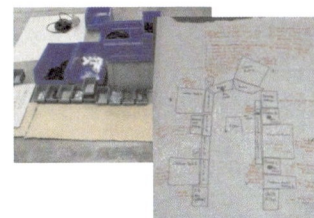

- Simulate, where necessary, the cell with full-size cardboard cut-outs
- Cell implementation
- Set performance and relative goals for the cell designed
- Establish a continuous improvement approach … (*perfection*)

# Cell Design

## Practical tips during cell design and implementation:

- Tools and materials should be located where needed ("Point Of Use")
- During cell design think about the reduction of walking distance
- Reduce unnecessary movements
- Materials must be in front of the workstation
- Use "Safety first" approach (the workplace must be safe and ergonomic)
- Materials supply should be simple and immediate
- Use 5S and Visual Management

# Cell Design

## Example of cell and cell management:

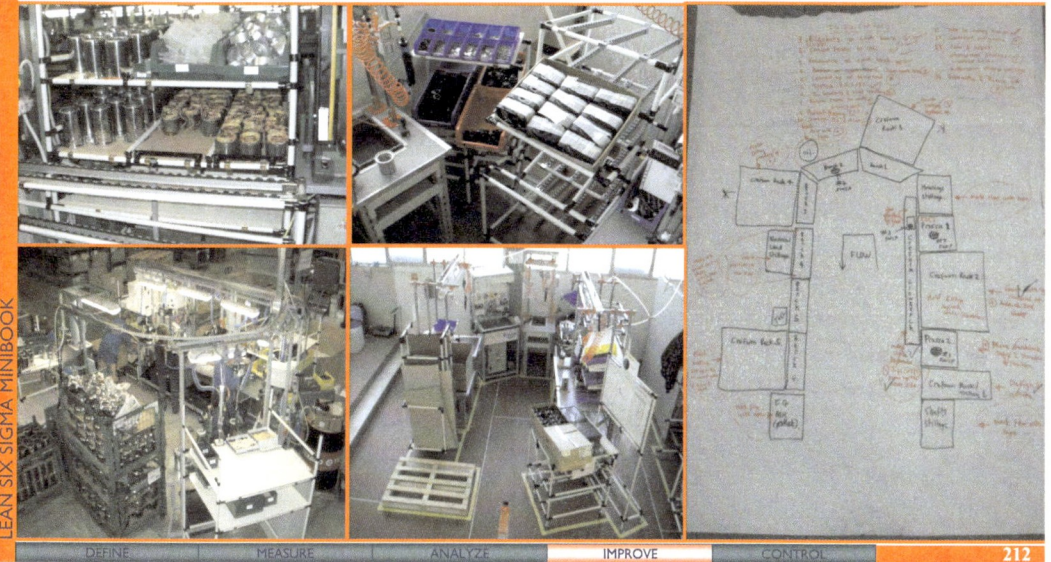

# SMED - Single Minute Exchange of Die

## Objective:

- The Single Minute Exchange of Die (SMED) is a method that aims to reduce the changeover time of machine equipment, or in general a production/service process

## Definition:

- **Changeover time**: it is the time required to prepare a device, machine, process, or system from the <u>last piece</u> of the previous batch to the <u>first good piece</u> of the next batch

# SMED - Single Minute Exchange of Die

When to use it:

- **Capacity problem:** it is necessary to reduce the changeover time in order to gain time available for production:

| Before | PRODUCTION | CHANGEOVER | PRODUCTION |
|---|---|---|---|

Time useful for more production

| After | PRODUCTION | CHANGEOVER | | PRODUCTION |
|---|---|---|---|---|

- **Need flexibility:** if the changeover time decreases the batch size can be smaller and consequently the flexibility of production can increase:

| Before | PRODUCT A | CHANGEOVER | PRODUCT B |
|---|---|---|---|

The mix of products at the end of the available time is bigger

| After | PRODUCT A | CHANGEOVER | PRODUCT B | CHANGEOVER | PRODUCT C | CHANGEOVER |
|---|---|---|---|---|---|---|

LEAN SIX SIGMA MINIBOOK

# SMED - Single Minute Exchange of Die

How to perform a SMED activity:

1. Analyze actual procedure and elementary steps to realize the changeover (use a videotape if possible, especially in multi-person changeover)
2. Establish goals (WIP; batch size; changeover time reduction, etc.)
3. Apply the general procedure for set-up reduction (see next page)
4. Perform a test to validate the new set-up procedure
5. Identify new improvement opportunities
6. Create a new standard operating procedure

## CAUTION: Definition of external and internal activity

**EXTERNAL ACTIVITY:** Activity that can be performed when the machine/process is running
**INTERNAL ACTIVITY**: Activity that can be performed only when the machine/process is stopped

# SMED - Single Minute Exchange of Die

## How to apply the general procedure:

| | |
|---|---|
| Product A  **CHANGEOVER TIME**  Product B | **STEP 0** STARTING STANDARD PROCEDURE |
| Product A EE I I EEE I I I I EEEE I I E I EE Product B | **STEP 1** ANALYZE IN THE CHANGEOVER STEPS WHICH ACTIVITIES CAN BE CONSIDERED INTERNAL AND WHICH EXTERNAL |
| Product A  I I I I I I I I I I  Product B  EEEEEEE  EEEEE | **STEP 2** MOVE THE EXTERNAL ACTIVITY BEFORE OR AFTER THE CHANGEOVER TIME |
| Product A  I I I I I I I I I I  Product B  EEEEEEE  EEEEE | **STEP 3** VERIFY IF IT IS POSSIBLE TO TRANSFORM SOME INTERNAL ACTIVITIES INTO EXTERNAL ACTIVITIES |
| Product A  I I I I I I  Product B  EEEEEEE  EEEEEEEE | **STEP 4** TRANSFORM THE INTERNAL ACTIVITIES INTO EXTERNAL AND MOVE OUTSIDE THE CHANGEOVER |
| Product A  I I I I I I  Product B  EEEEEEE  EEEEEEEE | **STEP 5** OPTIMIZED INTERNAL ACTIVITY (simplify, reduce, eliminate adjustments) |
| Product A  I I I  Product B  EEEE  EEEEEE | **STEP 6** REDUCE THE EXTERNAL ACTIVITIES |

New optimized Changeover Procedure

# SMED - Single Minute Exchange of Die

## How to collect data according to changeover activities:

|  |  | Observation Date | 15/04/2010 |  |  |  |  |
|---|---|---|---|---|---|---|---|
|  |  | Company | XYZ Company |  |  |  |  |
|  |  | Place | XXX |  |  |  |  |
|  |  | Kaizen | SMED |  |  |  |  |
|  |  | Kaizen Leader | Mr. Pink |  |  |  |  |
|  |  | Kaizen Team | Mr. Yellow; Mrs.Pink; Mr Grey; Ms. Orange |  |  |  |  |
|  |  | Last update | 15/04/2010 |  |  |  |  |

### Changeover Observation Form - 2 Operators

| N. | Activity | Who | I / E | Activity | Who | I / E | Time | Tool | Possible ideas or Improvements |
|---|---|---|---|---|---|---|---|---|---|
| 1 | MACHINE STOP | Op.1 | Internal | MACHINE STOP |  |  | 0' 00" |  |  |
| 2 | Open gate | Op.1 | Internal |  |  |  | 0' 50" |  |  |
| 3 | Cut coils | Op.1 | Internal |  |  |  | 1' 14" | Scissor | Which is the right position for scissors? (EHS)_5S |
| 4 | Passes behind | Op.1 | External |  |  |  | 1' 22" |  |  |
| 5 | Set the machine in manual configuration | Op.1 | Internal |  |  |  | 1' 31" |  |  |
| 6 | Rewinds coils | Op.1 | External |  |  |  | 3' 24" |  |  |
| 7 | Move the reel in order to put on the cradle | Op.1 | Internal |  |  |  | 3' 40" |  |  |
| 8 | He goes to the desk to take the key | Op.1 | External |  |  |  | 4' 48" |  |  |
| 9 | Raise the cradle (not high enough) | Op.1 | Internal |  |  |  | 6' 40" | Key | The thickness should be placed first on |
| 10 | Try to download the coil on the cradle | Op.1 | Internal | Takes the forklift | Op.2 | External | 7' 15" |  |  |
| 10 | Removes the tail of coil | Op.1 | External | Lifts fork | Op.2 | External | 8' 28" |  |  |
| 11 | Parades reel | Op.1 | Internal | He goes to pick up the new coil near the crane | Op.2 | External | 9' 14" |  |  |
| 12 | Puts the thickness on the cadle | Op.1 | Internal | Descends from the forklift | Op.2 | External | 11' 11" |  |  |
| 13 | Close the gate | Op.1 | Internal | Goes to the crane and take remote control | Op.2 | External | 11' 14" |  | NVA |
| 14 | Restart the machine | Op.1 | Internal | Guides the crane | Op.2 | External | 11' 40" |  |  |
| 15 | Molding of the last remaining pieces of the coil | Op.1 | Internal | Leads the hook near the new coil | Op.2 | External | 12' 23" |  |  |
| 16 | He puts the pieces on the pallet | Op.1 | Internal | He puts the hook on coil storaged on the cradle | Op.2 | External | 12' 45" |  |  |
| 17 | Removed ground plane from the press | Op.1 | Internal | Positioning with crane of the coil in the middle of the way | Op.2 | External | 13' 05" |  |  |
| 18 | Puts it next to the molds supermarket | Op.1 | Internal | Turns the coil for correct positioning | Op.2 | External | 14' 00" |  |  |
| ... | ... |  |  |  |  |  | ... | ... | ... |

# SMED - Single Minute Exchange of Die

Practical tips to improve changeover time performance:

- Use checklist to identify all the material necessary to perform the changeover activities
- Prepare all the raw materials before stopping the machine
- Check the raw materials to avoid placing the wrong one
- Try to plan the set-up in order to reduce the activity between the previous batch and the next one
- Try to perform the pre-heating of molds before machine stoppage in order to reduce scraps because of incorrect temperature of the machine
- Use, "visual changeover" where possible
- Try to standardize size of screws and bolts, height of dies, etc. as much as possible
- Use quick lock and quick release systems
- If possible, use "before and after" approach, all material in line approach, kit management approach or product family approach

# SMED - Single Minute Exchange of Die

Typical proportions of the activities in a changeover process:

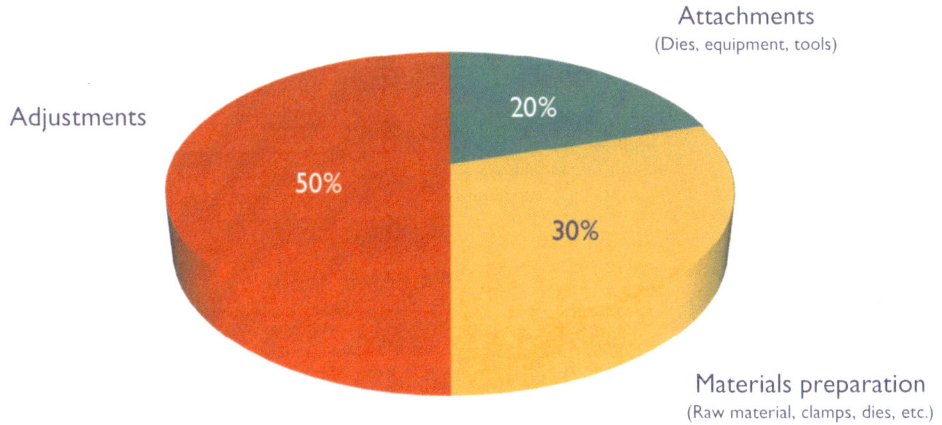

Attachments
(Dies, equipment, tools)

Adjustments

50%

20%

30%

Materials preparation
(Raw material, clamps, dies, etc.)

# SMED - Single Minute Exchange of Die

## Changeover optimization examples:

Use Visual Management to identify the right equipment for the changeover

All the tools necessary for the set-up must be prepared before changeover starts

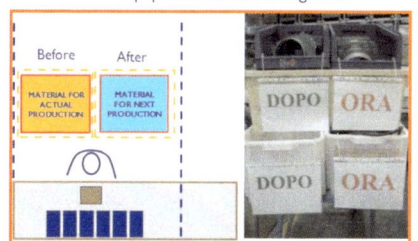

With before and after material management approach the production is already prepared for next batch

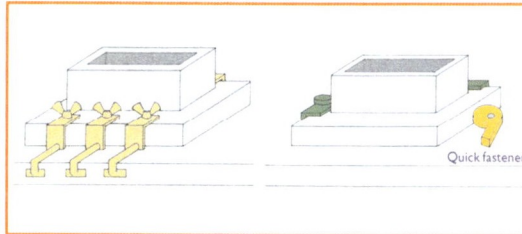

Use quick fastener system to reduce the attachment time as much as possible

# Kanban

## Objective:

- Kanban[G] means 'signboard'. Kanban can be used in many applications in various processes. It is primarily used as an instruction mechanism that controls the production, movement of goods, material, or parts, or jobs. For example, in controlling production, Kanban will tell you what to produce, when to produce, and how much to produce

## Benefits of Kanban:

- It is a basic tool for the pull system in production and supply processes
- It effectively reduces the inventory level of work in process (WIP), thus reducing wastes
- It reacts effectively with fluctuating demands from customers or downstream
- It is used to connect two processes with very different lead times
- It is a "visual management" system

# Kanban

## Withdrawal Kanban and Production Kanban:

Withdrawal Kanban: is used to order supplies/materials or command movement of process material/parts/semi-finished goods flows

Production Kanban: is used to control production, with detailed command including detailed information relating to quantity, type, destination etc.

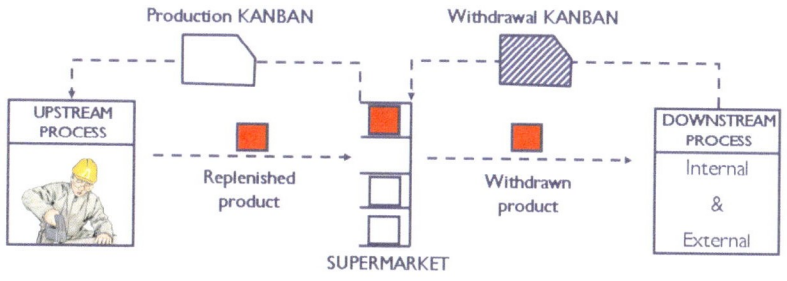

# Kanban

## Types of Withdrawal Kanbans:

- **Customer Kanban**
  - Transfers material from plant to customer
- **Move Kanban**
  - Transfers material between work processes
- **Supplier Kanban**
  - Pulls material from supplier to plant

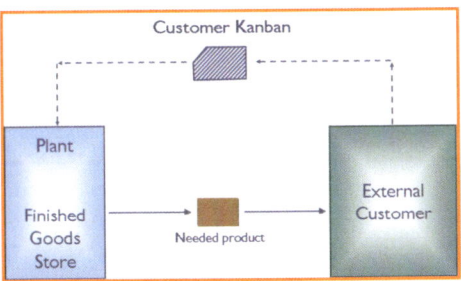

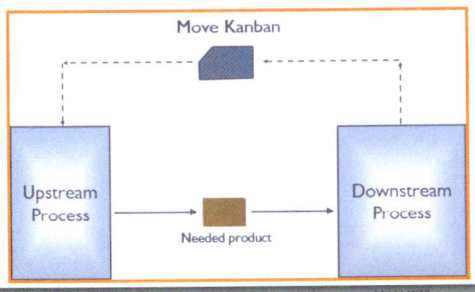

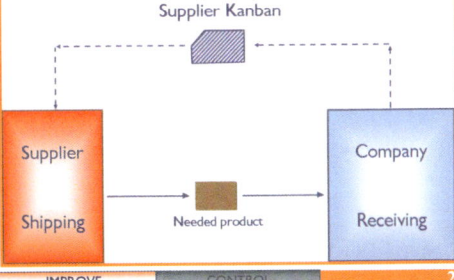

# Kanban

## Types of Production Kanban:

- **Production Kanban Card**
  - Gives instructions how to make one piece or one container
- **Trigger Kanban**
  - Instructs us to produce one batch
  - Used when the processes require set-up. The Kanban authorizes production only when the number of units to replenish is equal to the "Economic Order Quantity"

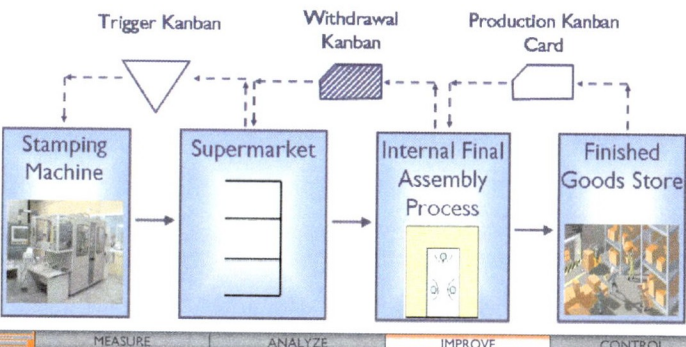

# Kanban

## Kanban sizing (analytical approach):

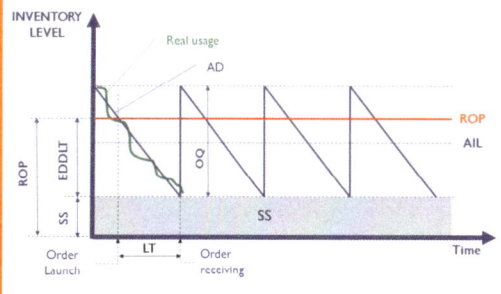

- **EDDLT** = Expected Demand During Lead Time (n° pcs)
- **AD** = Average Demand per day (n° pcs/day)
- **OQ** = Order Quantity (n° pcs)
- **ROP** = Reorder Point (n° pcs)
- **AIL** = Average Inventory Level (n° pcs)

$$AIL = OQ / 2 + SS$$

- **LT** = Total Lead Time (Production LT + Delivery LT (Days ))
- **SS** = Safety Stock (n° pcs)

$$SS = n° \ extra \ days \times AD$$

- **Q** = Size of container (n° pcs)

### SOME PRACTICAL TIPS:

- The minimun number of kanban is 2
- The Kanban number must be increased to one unit if the process starts with one empty bin
- If the bin is at the point of usage the Kanban number must be increased by one unit

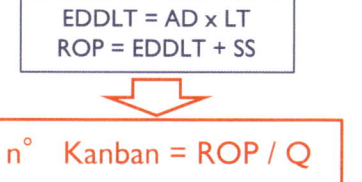

$$EDDLT = AD \times LT$$
$$ROP = EDDLT + SS$$

$$n° \ Kanban = ROP / Q$$

# Kanban

How to implement Kanban Process:

- Perform Kanban sizing
- Realize physical Kanban (bins, cards, etc.)
- Organize *supermarket*
- Implement FIFO mechanism

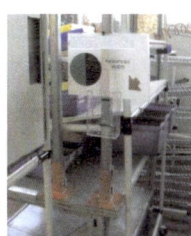

(Example kanban collection point)

- Use "Visual Management approach" (blackboard, labels, colored bins, etc.)
- Organize point of collection
- Train the people involved in kanban process
- Implement Kanban performance dashboard (Inventory level; Number of stock shortages, etc.)

---

### CAUTION:
Don't forget to start from products with high inventory turn index

---

# Kanban

## Kanban Card Designs:

### Customer Kanban

| Supplier | Quantity | Supplier No. | Card No. |
|---|---|---|---|
| Kanban Signal No. | Container Type | Part No. | Description |
| Storage Location | | Storage Address | |

### Move Kanban

| STORE ADDRESS | | WORK UNIT ADDRESS | |
|---|---|---|---|
| SUPPLIER NAME | Kanban Signal No. | | WORK UNIT NAME |
| PART No. | | QUANTITY | CARD # |

### Supplier Kanban

| Supplier Name | | Kanban Signal Number | |
|---|---|---|---|
| Part Number | | Quantity | Card No. |
| Part Name | | | |
| Customer | | Store Address | Work Unit Address |

### Production Kanban Card

| RECEIVING ADDRESS | | |
|---|---|---|
| Kanban No. | WORK UNIT NAME | |
| PART No. | QUANTITY | CARD # |

### Trigger Kanban

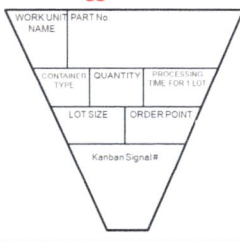

# Heijunka

## What is Heijunka?

- Heijunka[(G)] can be defined as "pursuit of even distribution of production volume and production mix over time"

## Why Heijunka is beneficial?

- Customer demands for products are uneven and non smooth in nature. However, the production processes work well at an even and smooth pace, Heijunka can effectively convert uneven demands into even ones and predictable production process by leveling production volume and mix
- Heijunka can be used in combination with other Lean tools, such as Kanban, and SMED to create smooth value flow

## Components of Heijunka:

1. Production leveling
2. Product mix leveling
3. Heijunka box

# Heijunka

## Why Production Volume Leveling?

- Fluctuations in customer demand often create greater disturbances in upper stream processes and supply chains
- Chasing variations in demand with fluctuation in production often causes: alternating overtime and idle time, quality problems, higher costs and stressed out workers

## What is Production volume Leveling?

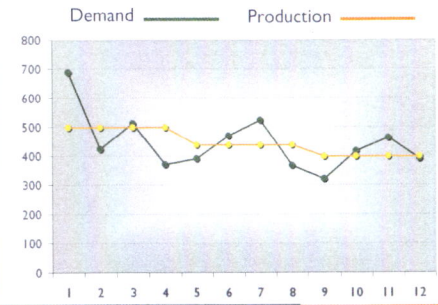

| Week | Demand |
|------|--------|
| 1 | 686 |
| 2 | 422 |
| 3 | 511 |
| 4 | 369 |
| 5 | 390 |
| 6 | 468 |
| 7 | 523 |
| 8 | 367 |
| 9 | 320 |
| 10 | 416 |
| 11 | 462 |
| 12 | 390 |

1988 / 4 = 497

1748 / 4 = 437

1588 / 4 = 397

| Weekly leveled production |
|---------------------------|
| 497 |
| 497 |
| 497 |
| 497 |
| 437 |
| 437 |
| 437 |
| 437 |
| 397 |
| 397 |
| 397 |
| 397 |

LEAN SIX SIGMA MINIBOOK

# Heijunka

## Why Product Mix Leveling?

- Large batches of the same product may reduce number of changeovers, but it usually causes long lead time and high inventory level
- Variations in daily work loads will lead to uneven work pace, possibly creating excessive idle time, overtime, and reducing quality
- A daily production schedule with stable product mix and stable work load will make a stable production process

## What is Product Mix Leveling?

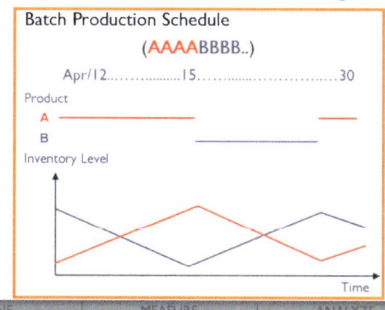

Batch Production Schedule
(AAAABBBB..)

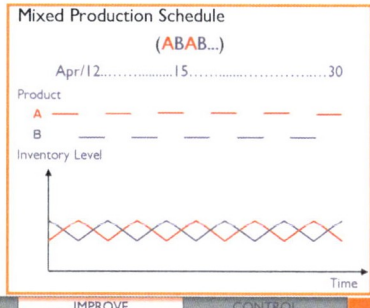

Mixed Production Schedule
(ABAB...)

# Heijunka

## A Product Mix Leveling Example:

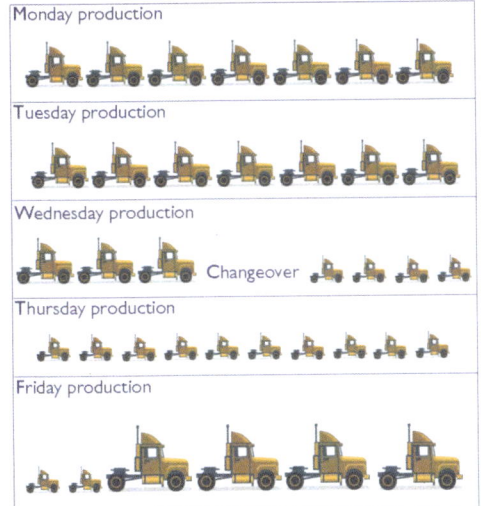

Batches, uneven daily loads

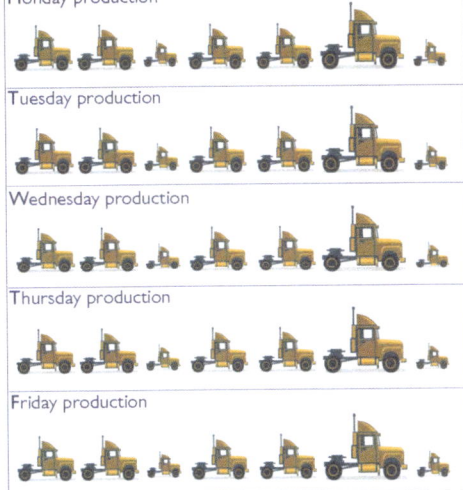

Stable mix, stable daily loads

# Heijunka

## What is Heijunka box?

- The Heijunka box is a visual scheduling tool for Heijunka application, both production volume leveling and product mix leveling are visually displayed

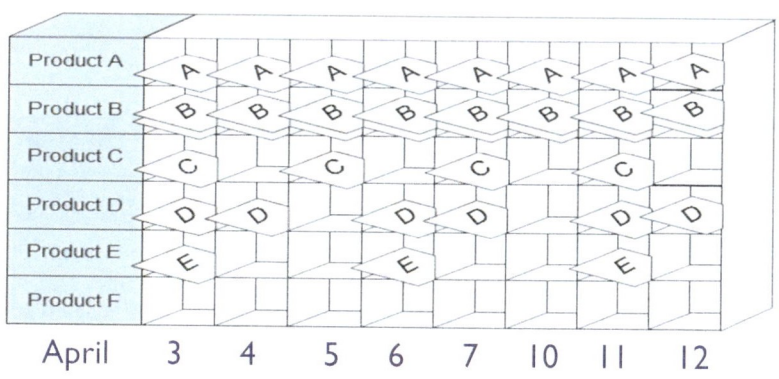

# TPM - Total Productive Maintenance

## Objective:

- Total Productive Maintenance (TPM) is a methodology focused on the technical aspects of manufacturing processes. It aims to increase plant and equipment productive performances through employees' empowerment and skills. For that reason the real owner of the methodology is not only maintenance but all the production system

## TPM pillars:

- Total Productive Maintenance is based on 8* pillars

  - Focus improvement
  - Autonomous Maintenance
  - Planned Maintenance
  - Quality maintenance

  - Early equipment management
  - Office TPM
  - Education and training
  - Safety, Health and environment

\* The number of pillars can change from company to company

# TPM - Total Productive Maintenance

Overview:

- The TPM methodology is represented by a temple where at the base there are two important points: lean waste culture and 5S approach (don't forget to use them everytime you look at the process optimization)

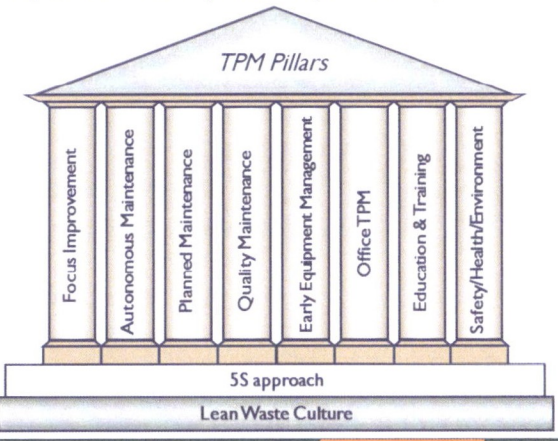

LEAN SIX SIGMA MINIBOOK

# TPM - Total Productive Maintenance -

## What is the meaning of the 8 pillars?

### FOCUS IMPROVEMENT

Focus Improvement, or Kobetsu Kaizen, aims to maximize the overall system efficiency (generally measured with OEE index) through the elimination of equipment/process losses. It is one of the most important activities because of its rapid implementation and impact dimension. This pillar is strictly related to all other pillars

### PLANNED MAINTENANCE

Planned maintenance aims to move away from reactive maintenance to a proactive approach (planned maintenance). The activity is complementary to autonomous maintenance and generally requires the maintenance team. It aims to improve the technique and practice of maintenance, due to the increase of skills and ability to use diagnostic techniques. The ideal goal is "Zero breakdowns"

### AUTONOMOUS MAINTENANCE

Autonomous maintenance is the involvement of production workers in machine/equipment management through the transfer of some activity like daily inspection, lubrication, cleaning, minor repairs, etc. typical of a maintenance team. The operator is the expert on the equipment and for this reason he is primarily responsible for machine care. Continuous improvement is the basis of this pillar

### QUALITY MAINTENANCE

Quality Maintenance aims to create and maintain the conditions of the process from which the products come in order to ensure the quality level required by the customer. The starting point is not the product but the process control and consequently the management of its conditions (eg. pressure, temperature, density, flow rate, etc.) to maintain the desired result

# TPM - Total Productive Maintenance -

## What is the meaning of the 8 pillars?

### EARLY EQUIPMENT MANAGEMENT

Early Equipment Management is a structured process focusing on reducing the complexity associated with the operation and maintenance of equipment. It helps to extend the principles of Lean to the design and manufacture of equipment

### EDUCATION & TRAINING

Education and training pillar is the basis of TPM. Its goal is to ensure that all people involved in TPM have all the skills necessary to support the change. Improving knowledge and skills of operators will not only increase their operational abilities, but encourage pride in their work, with great benefits to all. The two main ways to reach this are "on the job training" and "self-development"

### OFFICE TPM

The Office TPM pillar aims to involve the non-productive departments to focus on better plant performance. The pillar helps administrative functions in defining their goals to support the TPM process in the production area

### SAFETY, HEALTH, ENVIRONMENT

This pillar is really important for all company systems. TPM doesn't only mean more efficiency of equipment or less defects, but also eliminates all problems related to safety and environment. The main goal is to create a workplace that is organized, safe and environmentally oriented

# Priority Matrix

## Objective:

- *Priority Matrix* can be used to quantify the degree of correlation between Input and Output variables

- In addition, it is often used to identify the best solutions according to criteria, weighting them appropriately (e.g. it can be useful when consensus is not reached in making a group decision)

- In a Lean Six Sigma project, this method can be used in Measure Phase to identify linkage between input and output variables. It can be used in Improve Phase to identify key variables/solutions to be implemented in order to solve the problem

# Priority Matrix

**How** to Build a Priority Matrix*:

1. Identify criteria for evaluating different solutions
2. Team members assign weight to all criteria. For each member, all weight assigned to criteria should added up to 1 (0 is allowed in weighting)

|  | Marco | Stefano | Claudio | Mirko |  |
|---|---|---|---|---|---|
| Criterion 1 |  | 0,1 | 0,2 |  | 0,3 |
| Criterion 2 | 0,25 | 0,45 | 0,4 | 0,6 | 1,7 |
| Criterion 3 |  | 0,2 | 0,1 |  | 0,3 |
| Criterion 4 | 0,25 | 0,25 | 0,2 | 0,4 | 1,1 |
| Criterion 5 | 0,5 |  | 0,1 |  | 0,6 |
|  | 1 | 1 | 1 | 1 |  |

* The example is related to a case of selecting the optimal solution in which four people suggest the selection criteria and the weight to be associated with the solutions to implement

# Priority Matrix

3. Sum the values for each criteria to get the total weight

|  | Marco | Stefano | Claudio | Mirko |  |
|---|---|---|---|---|---|
| Criterion 1 |  | 0,1 | 0,2 |  | **0,3** |
| Criterion 2 | 0,25 | 0,45 | 0,4 | 0,6 | **1,7** |
| Criterion 3 |  | 0,2 | 0,1 |  | **0,3** |
| Criterion 4 | 0,25 | 0,25 | 0,2 | 0,4 | **1,1** |
| Criterion 5 | 0,5 |  | 0,1 |  | **0,6** |
|  | 1 | 1 | 1 | 1 |  |

|  | Evaluation criterion 1 | Evaluation criterion 2 | Evaluation criterion 3 | Evaluation criterion 4 | Evaluation criterion 5 |  |
|---|---|---|---|---|---|---|
| Weights | 0,3 | 1,7 | 0,3 | 1,1 | 0,6 | TOTAL |
| Solution A |  |  |  |  |  |  |
| Solution B |  |  |  |  |  |  |
| Solution C |  |  |  |  |  |  |
| Solution D |  |  |  |  |  |  |

# Priority Matrix

4. Each evaluator assigns a score of 1, 3 or 5 to rate the ability of each solution to satisfy a criterion

5. Sum the scores from all evaluators and multiply the result by the weight assigned for each criteria, and fill out the following matrix

| | Evaluation criterion 1 | Evaluation criterion 2 | Evaluation criterion 3 | Evaluation criterion 4 | Evaluation criterion 5 | |
|---|---|---|---|---|---|---|
| Weights | 0,3 | 1,7 | 0,3 | 1,1 | 0,6 | TOTAL |
| Solution A | (5+5+5+3)x0,3 | (1+3+3+3)x1,7 | (5+3+3+1)x0,3 | (5+3+5+3)x1,1 | (5+1+1+3)x0,6 | |
| Solution B | (3+3+1+1)x0,3 | (1+1+1+3)x1,7 | (5+3+1+1)x0,3 | (5+5+5+3)x1,1 | (5+3+3+1)x0,6 | |
| Solution C | (5+1+1+1)x0,3 | (5+5+1+1)x1,7 | (5+5+5+1)x0,3 | (1+1+1+1)x1,1 | (5+3+1+1)x0,6 | |
| Solution D | (5+3+3+1)x0,3 | (5+1+1+1)x1,7 | (5+3+1+1)x0,3 | (5+5+5+3)x1,1 | (3+1+1+3)x0,6 | |

# Priority Matrix

6. Sum the values for all rows to get the total score for each solution. The solution with the highest score means that this solution is the best one in terms of ability to fairly satisfy all weighted criteria

| | Evaluation criterion 1 | Evaluation criterion 2 | Evaluation criterion 3 | Evaluation criterion 4 | Evaluation criterion 5 | |
|---|---|---|---|---|---|---|
| Weights | 0,3 | 1,7 | 0,3 | 1,1 | 0,6 | TOTAL |
| Solution A | 5,4 | 17 | 3,6 | 17,6 | 6 | 49,6  A |
| Solution B | 2,4 | 10,2 | 3 | 19,8 | 7,2 | 42,6 |
| Solution C | 2,4 | 20,4 | 4,8 | 4,4 | 6 | 38 |
| Solution D | 3,6 | 13,6 | 3 | 19,8 | 4,8 | 44,8 |

A This is the solution that has the highest impact based on weighted criteria

LEAN SIX SIGMA MINIBOOK

# FMEA

Objective:

- FMEA[G] (*Failure Modes and Effects Analysis*) is used to identify a detailed list of failure modes of a product or process and their corresponding causes and then rate them with severity level, likelihood of occurrence and detection in order to manage system risk

- In Lean Six Sigma projects, FMEA can be used as a systematic method to link inputs with outputs, assign priority levels and degree of relationship, or to assess risk associated with different solutions to be implemented. For these reasons this technique can be applied to different stages of DMAIC

# FMEA

RPN (*Risk Priority Number*) is a risk management index that is the product of the following 3 values:

- O: *Occurrence* or probability of occurrence, which relates to causes
- S: *Severity* of the failure effect
- D: *Detection*, Chance of detecting the failure before it happens

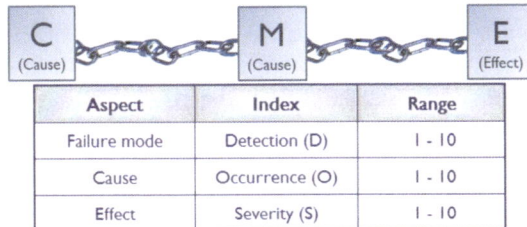

| Aspect | Index | Range |
|--------|-------|-------|
| Failure mode | Detection (D) | 1 - 10 |
| Cause | Occurrence (O) | 1 - 10 |
| Effect | Severity (S) | 1 - 10 |

$$RPN = O \times S \times D$$

## Rule of Thumb:

Corrective action must be taken if RPN value exceeds a threshold (e.g., 100) or severity exceeds 8

# FMEA

1. Identify potential failure modes
2. Identify potential failure effect and its severity (Severity)
3. Identify the failure causes and the likelihood of occurrence (Occurrence)
4. Identify the effectiveness of current system to detect failure (Detection)
5. Multiply the values (S, O, D) to determine the risk level of each failure mode (RPN)
6. Identify the corrective actions for the failure modes that have high RPN value (greater than the threshold value) or when Severity value is more than 8 (9 or 10)
7. Recalculate RPN after improvement. Check the improvement until both RPN and Severity values are acceptable

# FMEA

| Feature | O Value | Likelihood of Occurrence | Feature | S Value |
|---|---|---|---|---|
| Very low likelihood | 1 | <1/100.000 | Slight | 1 |
| | | | | 2 |
| Low likelihood | 2 | <1/20.000 | Slightly important | 3 |
| | 3 | <1/10.000 | | 4 |
| Medium likelihood | 4 | <1/2.000 | Moderate | 5 |
| | 5 | <1/1.000 | | 6 |
| | 6 | <1/200 | | |
| High likelihood | 7 | <1/100 | High | 7 |
| | 8 | <1/20 | | 8 |
| Very hight likelihood | 9 | <1/10 | High (Safety problems) | 9 |
| | 10 | >1/10 | | 10 |

| Feature | D Value | Likelihood of Detection |
|---|---|---|
| Very high | 1 | =100% |
| | 2 | > 99% |
| High | 3 | >95% |
| | 4 | >90% |
| Medium | 5 | >50% |
| | 6 | >20% |
| Low | 7 | >10% |
| | 8 | >5% |
| Very low | 9 | >1% |
| | 10 | =0% |

C = Cause
M = Failure Mode
E = Effect

(O) 1-10

C

(D) 1-10

M

(S) 1-10

E

$$RPN = O \times S \times D$$

# FMEA

## Example of FMEA application:

| PART / PROCESS STEP | MODE | EFFECT | CAUSE | PARAMETERS | | | RPN (Risk Priority Number) | CORRECTIVE ACTIONS | Responsability | PARAMETER S After corrective actions | | | RPN (Risk Priority Number) |
|---|---|---|---|---|---|---|---|---|---|---|---|---|---|
| | | | | O | D | S | | | | O | D | S | |
| Step 1 | Mode 1.1 | Effect 1.1 | Cause 1.1 | 3 | 5 | 6 | 90 | | | | | | 0 |
| | | | Cause 1.2 | 2 | 5 | 6 | 60 | | | | | | 0 |
| | Mode 1.2 | | Cause 1.3 | 8 | 3 | 6 | 144 | Action 1.3 | P.White | 3 | 3 | 8 | 72 |
| | | | Cause 1.4 | 5 | 3 | 6 | 90 | | | | | | 0 |
| Step2 | Mode 2.1 | Effect 2.1 | Cause 2.1 | 6 | 8 | 3 | 144 | Action 2.1 | G. Blue | 6 | 5 | 3 | 90 |
| | | | Cause 2.2 | 5 | 8 | 3 | 120 | Action 2.2 | G. Blue | 5 | 5 | 3 | 75 |
| | Mode 2.2 | Effect 2.2 | Cause 2.3 | 3 | 9 | 4 | 108 | Action 2.3 | G. Blue | 3 | 2 | 4 | 24 |
| Step 3 | Mode 3.1 | Effect 3.1 | Cause 3.1 | 2 | 2 | 10 | 40 | Action 3.1 | V. Black | 7 | 3 | 4 | 84 |
| | Mode 3.2 | Effect 3.2 | Cause 3.3 | 3 | 2 | 7 | 42 | | | | | | 0 |
| | Mode 3.3 | | | 3 | 1 | 7 | 21 | | | | | | 0 |
| | | | Total risk index = | | | | 859 | Total risk index (After) = | | | | | 345 |

# DOE

## Objective:

- DOE[G] (*Design Of Experiments*) is a methodology that builds, through well-planned experiments and analysis of the experimental results, the analytical model relating to the cause-effect relationship between input and output variables

## Fundamental Assumptions:

- Residuals (difference between actual response and model prediction) is normally distributed
- Residuals are independent of Xs (Input variables)
- Residuals are independent of predicted Ys (Fitted Value)
- Residuals are independent of time

## When to use it:

- DOE can be used in Analyze Phase to identify key variables and interactions that influence the output. DOE can also be used in the Improve Phase to identify best parameter settings (Inputs) to optimize the Output variable

# DOE

1. Define the problem

2. Select output (response) variable y

3. Select experimental factors (input variables), their levels and values

4. Select the experimental design in order to manage the number of tests. You can choose *Full Factorial Design*, which is a complete combination of all the levels of all factors, or *Fractional Factorial Design*, which is a selected portion of full factorial design. Fractional Factorial is economical but it lacks some information (some data analysis issues can be handled via *alias structure*, *resolution* and *confounding*)

5. Conduct experiments properly (randomization may be needed, proper data collection, etc.)

6. Analyze the experimental data

LEAN SIX SIGMA MINIBOOK

# DOE

Types of factorial experiments:

Replicates

General Full Factorial Design ⟶ Run = a × b × c × d × ... × n

Number of levels for factor A        Number of levels for Factor B

2 Level Full Factorial Design ⟶ Run = n × $2^k$

Number of factors

Replicates    Number of levels

2 Level Fractional Factorial Design ⟶ Run = n × $2^{k-q}$

Degree of fraction:

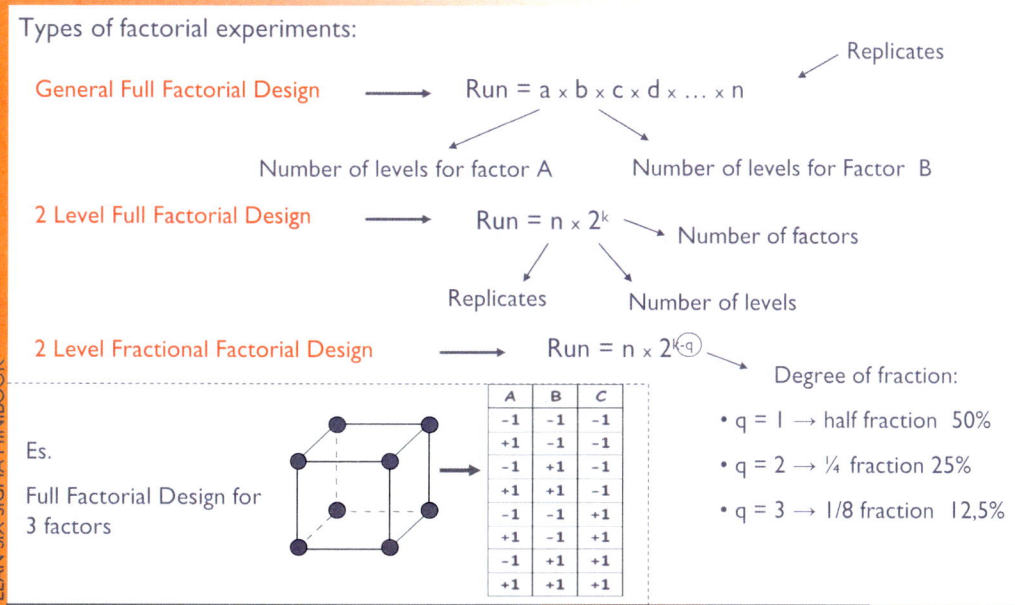

Es.

Full Factorial Design for 3 factors

| A | B | C |
|----|----|----|
| -1 | -1 | -1 |
| +1 | -1 | -1 |
| -1 | +1 | -1 |
| +1 | +1 | -1 |
| -1 | -1 | +1 |
| +1 | -1 | +1 |
| -1 | +1 | +1 |
| +1 | +1 | +1 |

- q = 1 → half fraction 50%
- q = 2 → ¼ fraction 25%
- q = 3 → 1/8 fraction 12,5%

# DOE

Create an experimental plan by MINITAB:
$\underline{S}$tat > $\underline{D}$OE > $\underline{F}$actorial > $\underline{C}$reate Factorial Design…

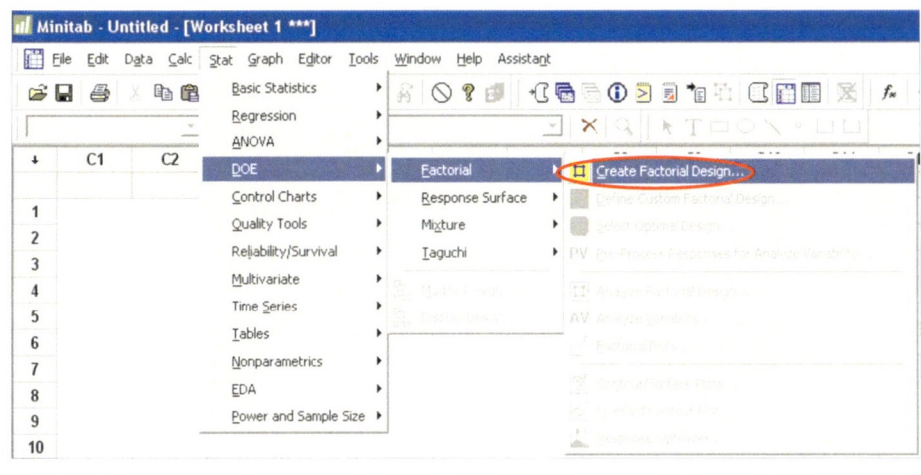

LEAN SIX SIGMA MINIBOOK

# DOE

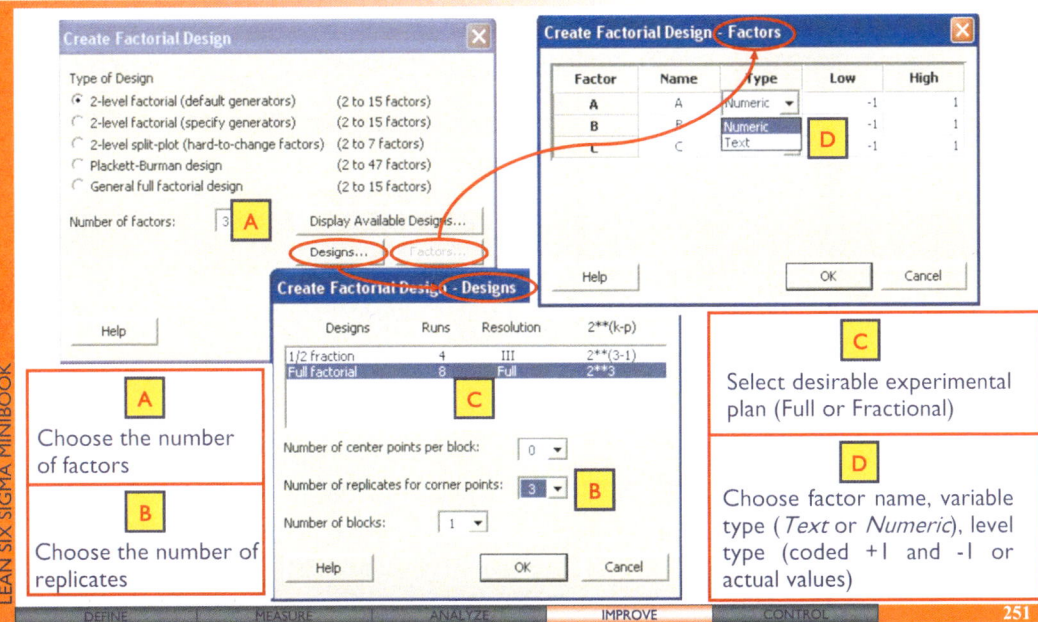

# DOE

## MINITAB: Output

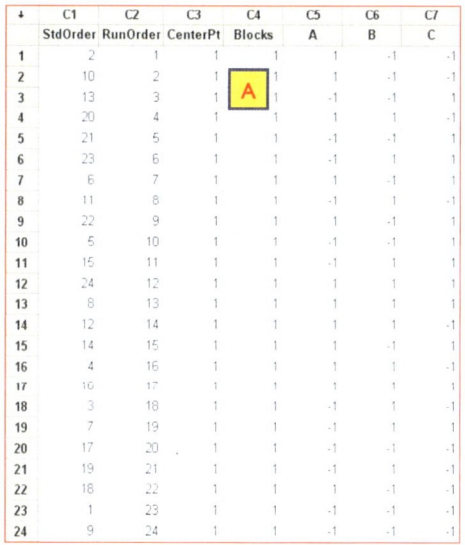

| ↓ | C1 StdOrder | C2 RunOrder | C3 CenterPt | C4 Blocks | C5 A | C6 B | C7 C |
|---|---|---|---|---|---|---|---|
| 1 | 2 | 1 | 1 | 1 | 1 | -1 | -1 |
| 2 | 10 | 2 | 1 | 1 | 1 | -1 | -1 |
| 3 | 13 | 3 | 1 | 1 | -1 | -1 | 1 |
| 4 | 20 | 4 | 1 | 1 | 1 | 1 | -1 |
| 5 | 21 | 5 | 1 | 1 | -1 | -1 | 1 |
| 6 | 23 | 6 | 1 | 1 | -1 | 1 | 1 |
| 7 | 6 | 7 | 1 | 1 | 1 | -1 | 1 |
| 8 | 11 | 8 | 1 | 1 | -1 | 1 | 1 |
| 9 | 22 | 9 | 1 | 1 | 1 | -1 | 1 |
| 10 | 5 | 10 | 1 | 1 | -1 | -1 | 1 |
| 11 | 15 | 11 | 1 | 1 | 1 | 1 | 1 |
| 12 | 24 | 12 | 1 | 1 | 1 | 1 | 1 |
| 13 | 8 | 13 | 1 | 1 | 1 | 1 | 1 |
| 14 | 12 | 14 | 1 | 1 | 1 | 1 | -1 |
| 15 | 14 | 15 | 1 | 1 | 1 | -1 | 1 |
| 16 | 4 | 16 | 1 | 1 | 1 | 1 | -1 |
| 17 | 16 | 17 | 1 | 1 | 1 | 1 | 1 |
| 18 | 3 | 18 | 1 | 1 | -1 | 1 | 1 |
| 19 | 7 | 19 | 1 | 1 | -1 | 1 | 1 |
| 20 | 17 | 20 | 1 | 1 | -1 | -1 | -1 |
| 21 | 19 | 21 | 1 | 1 | -1 | 1 | -1 |
| 22 | 18 | 22 | 1 | 1 | -1 | 1 | -1 |
| 23 | 1 | 23 | 1 | 1 | -1 | -1 | -1 |
| 24 | 9 | 24 | 1 | 1 | -1 | -1 | -1 |

**Full Factorial Design**  B

| Factors: | 3 | Base Design: | 3; 8 |
|---|---|---|---|
| Runs: | 24 | Replicates: | 3 |
| **Blocks:** | **1** | **Center pts (total):** | **0** |

All terms are free from aliasing.

---

**A**

Experimental plan created

**B**

Experimental design information displayed in *session window*

LEAN SIX SIGMA MINIBOOK

# DOE

Manage statistical analysis plan

$\underline{S}$tat > $\underline{D}$OE > $\underline{F}$actorial > Analyze Factorial Design…

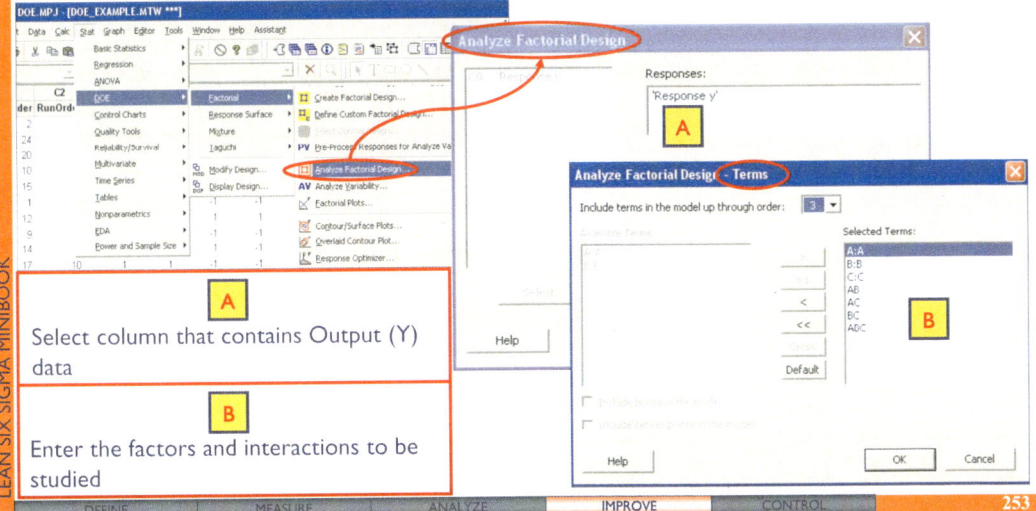

LEAN SIX SIGMA MINIBOOK

A

Select column that contains Output (Y) data

B

Enter the factors and interactions to be studied

## MINITAB: Output

**Factorial Fit: Response y versus A; B; C**

Estimated Effects and Coefficients for Response y (coded units)

| Term | Effect | Coef | SE Coef | T | P |
|---|---|---|---|---|---|
| Constant | | 0,3223 | 0,02912 | 11,07 | 0,000 |
| A | -0,2143 | -0,1071 | 0,02912 | -3,68 | 0,002 |
| B | -0,0107 | -0,0054 | 0,02912 | -0,18 | 0,858 |
| C | 0,2232 | 0,1116 | 0,02912 | 3,83 | 0,001 |
| A*B | -0,0268 | -0,0134 | 0,02912 | -0,46 | 0,652 |
| A*C | -0,2286 | -0,1143 | 0,02912 | -3,92 | 0,001 |
| B*C | -0,0786 | -0,0393 | 0,02912 | -1,35 | 0,196 |
| A*B*C | 0,0125 | 0,0062 | 0,02912 | 0,21 | 0,833 |

S = 0,142678  PRESS = 0,732857
R-Sq = 74,08%  R-Sq(pred) = 41,68%  R-Sq(adj) = 62,74%

Analysis of Variance for Response y (coded units)

| Source | DF | Seq SS | Adj SS | Adj MS | F | P |
|---|---|---|---|---|---|---|
| Main Effects | 3 | 0,57515 | 0,575147 | 0,191716 | 9,42 | 0,001 |
| A | 1 | 0,27551 | 0,275810 | 0,275810 | 13,53 | 0,002 |
| B | 1 | 0,00069 | 0,000689 | 0,000689 | 0,03 | 0,858 |
| C | 1 | 0,29895 | 0,298948 | 0,298948 | 14,69 | 0,001 |
| 2-Way Interactions | 3 | 0,35482 | 0,354815 | 0,118272 | 5,81 | 0,007 |
| A*B | 1 | 0,00430 | 0,004305 | 0,004305 | 0,21 | 0,652 |
| A*C | 1 | 0,31347 | 0,313469 | 0,313469 | 15,40 | 0,001 |
| B*C | 1 | 0,03704 | 0,037041 | 0,037041 | 1,82 | 0,196 |
| 3-Way Interactions | 1 | 0,00094 | 0,000937 | 0,000937 | 0,05 | 0,833 |
| A*B*C | 1 | 0,00094 | 0,000937 | 0,000937 | 0,05 | 0,833 |
| Residual Error | 16 | 0,32571 | 0,325714 | 0,020357 | | |
| Pure Error | 16 | 0,32571 | 0,325714 | 0,020357 | | |
| Total | 23 | 1,25661 | | | | |

### CAUTION:

Eliminate factors (one at a time) until all remaining ones are significant

**A**

For any main effect, two way, three way or higher order interaction, it is significant only if its P-Value < 0.05

**B**

Regression coefficients for the analytical model

**C**

Global importance of factors: if its P-Value is < 0.05 , then at least one single factor or interaction (2 way or 3 way, etc.) is significant

# DOE

## DOE Graphs

### Stat > DOE > Factorial > Factorial Plots…

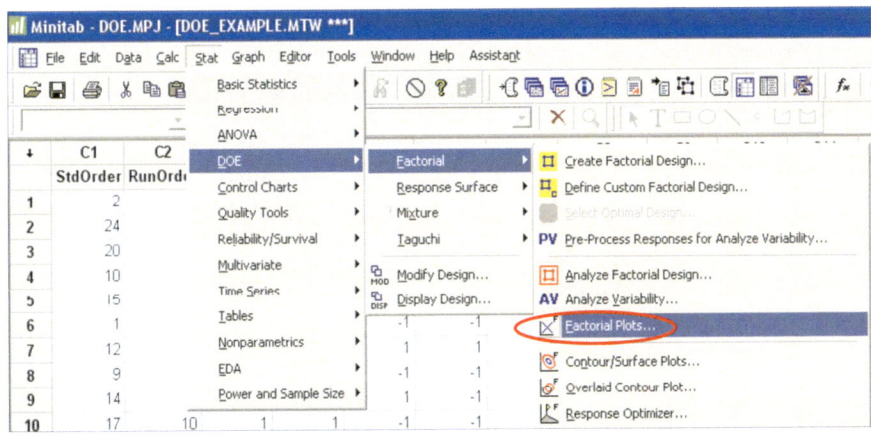

# DOE

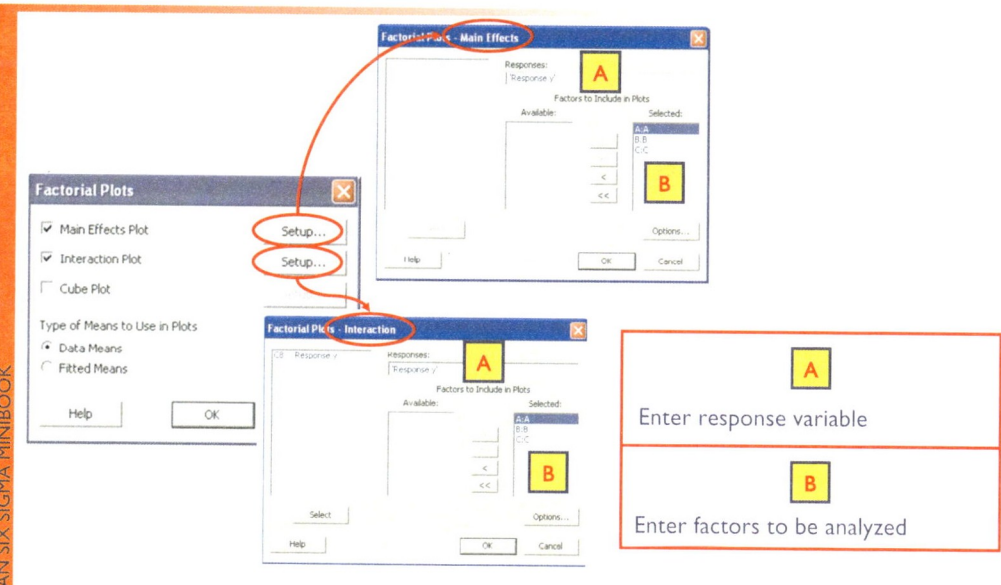

A — Enter response variable

B — Enter factors to be analyzed

# DOE

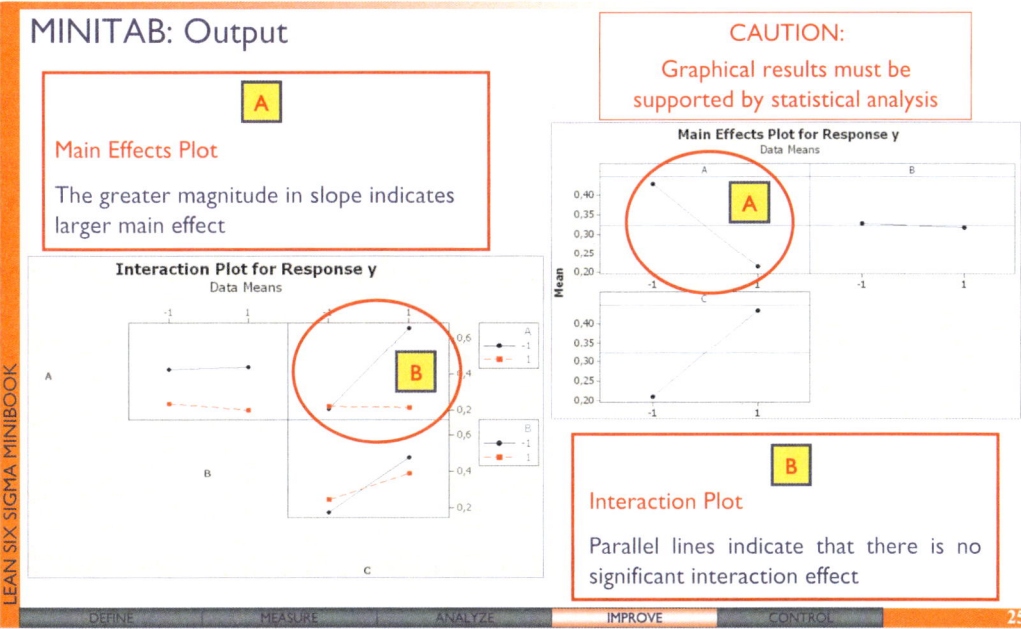

## MINITAB: Output

### A

**Main Effects Plot**

The greater magnitude in slope indicates larger main effect

**CAUTION:**

Graphical results must be supported by statistical analysis

### B

**Interaction Plot**

Parallel lines indicate that there is no significant interaction effect

# DOE: Assumptions

## Residual analysis

### <u>S</u>tat > <u>D</u>OE > <u>A</u>nalyze Factorial Design…(Graphs)

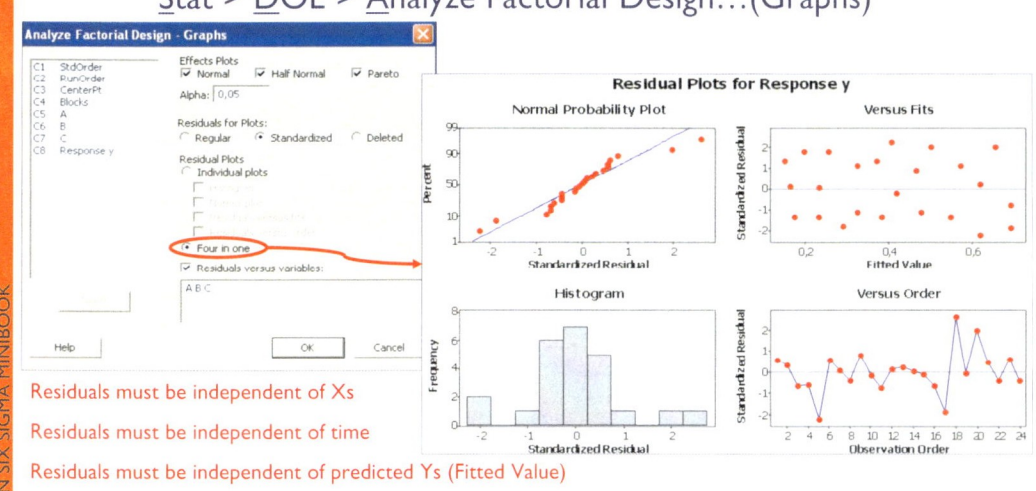

Residuals must be independent of Xs

Residuals must be independent of time

Residuals must be independent of predicted Ys (Fitted Value)

Residuals should be normally distributed

LEAN SIX SIGMA MINIBOOK

# CONTROL

LEAN SIX SIGMA MINIBOOK

Arcidiacono G., Calabrese C., Yang K.: Leading processes to lead companies: Lean Six Sigma.
DOI 10-1007/978-88-470-2492-2, © Springer-Verlag Italia 2012

# CONTROL

*Control* is the final step of Lean Six Sigma roadmap; the objective of this phase is to:

- Test the quality level because it is the result of previous Lean Six Sigma steps
- Validate the method and its effectiveness used in improvement
- Standardize the method if its effectiveness is proven
- Implement control plan to sustain the improved long term performance
- Use *visual management* and an *error proofing system* to maintain high level performance
- Verify the applicabilities and possible extensions of the method for other possible problems or company areas

# Control Chart

**Objective:**

- Control Charts are useful tools that can verify and monitor the stability of performance levels for manufacturing, transactional and service processes
- Control Charts are tools that can be used to identify "special causes' in the process

**Features:**

- **Common Cause**: The cause, random in nature and not related to any special event, is behind natural inherent variability shown in processes
- **Special Cause:** The cause is often associated with special events. The result of a Special Cause is that the process often shows a trend, seasonality or other non-random patterns

A process is "stable" if it is only influenced by random causes (common causes)

# Control Chart: Individuals

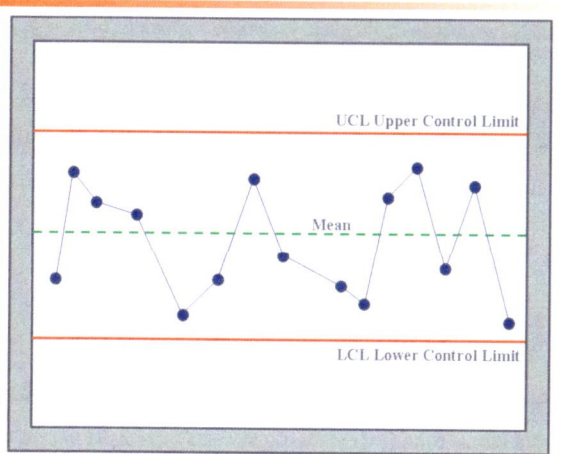

When to use it:

- When it is necessary to monitor individual continuous variables

$$UCL = \mu + L\sigma$$

$$CL = \mu$$

$$LCL = \mu - L\sigma$$

Formula for **control limits** calculation

- Control limits are specified by the **Upper Control Limit** (UCL) and **Lower Control Limit** (LCL), They determine the range for natural variation due to random causes. Any variation beyond UCL or LCL will be considered as 'out of control' and likely caused by a special event

# Control Chart: Individuals

MINITAB:

$\underline{S}$tat > $\underline{C}$ontrol Charts > Variables Charts for $\underline{I}$ndividuals…

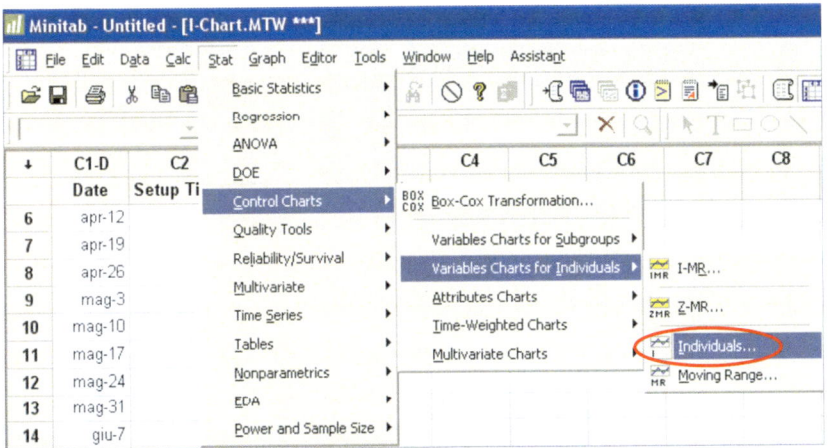

# Control Chart: Individuals

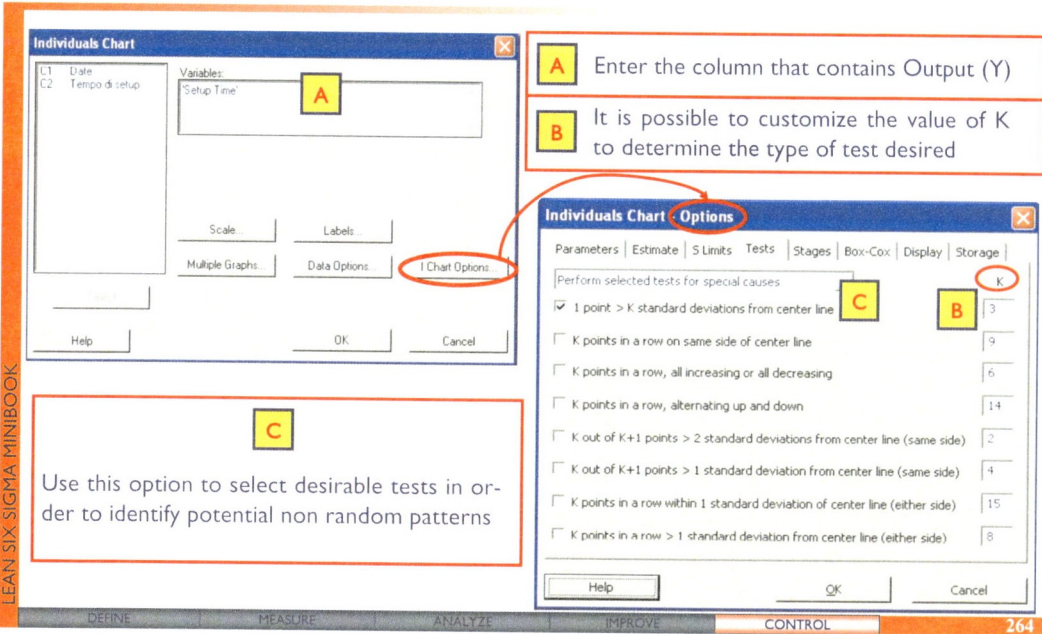

**A**   Enter the column that contains Output (Y)

**B**   It is possible to customize the value of K to determine the type of test desired

**C**   Use this option to select desirable tests in order to identify potential non random patterns

# Control Chart: Individuals

## MINITAB: Output

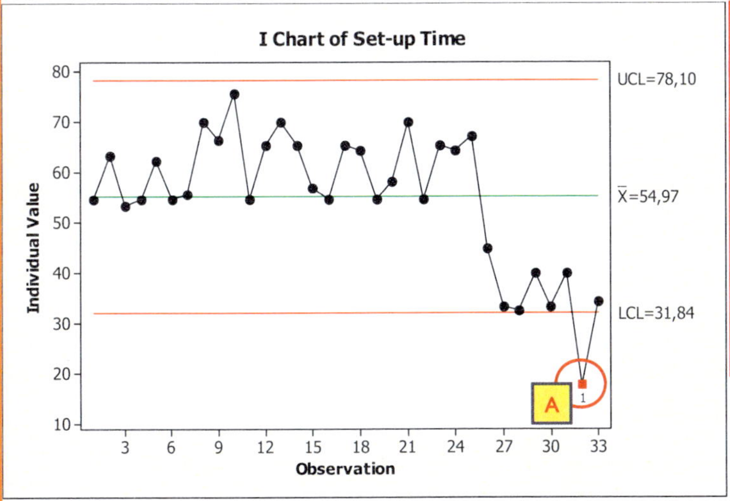

**I Chart of Set-up Time**

UCL=78,10

$\overline{X}$=54,97

LCL=31,84

A

**A**

A point outside control limit indicating the presence of special cause

It is necessary to investigate the causes that lead this out of control situation, in order to intervene and correct the problem

# Control Chart: Individuals

## Rules to determine out of control and non random trends

- One or more points out of control limits (usually 3-sigma lines)
- Two out of three points are beyond 2-sigma line (warning line) but usually within 3-sigma line
- Four out of five points fall beyond 1-sigma line from center
- Eight consecutive points fall on one side of centerline
- Six consecutive points are in ascending or descending order
- 15 consecutive points are in zone C (both above or below centerline)
- 14 consecutive points are alternating up and down (zig-zag)
- 8 consecutive points alternate around centerline, but none in zone C
- Non random patterns are observed

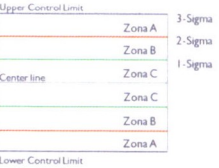

LEAN SIX SIGMA MINIBOOK

# Control Chart: Xbar-R

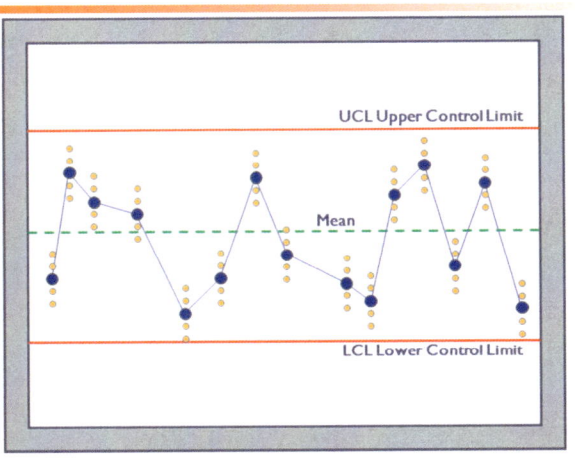

When to use it:

- It is necessary to monitor subgroups of continuous variables

- Xbar-R charts are often used in high volume production process monitoring

- Control limits are specified by the **Upper Control Limit** (UCL) and **Lower Control Limit** (LCL). They determine the range for natural variation due to random causes. Any variation beyond UCL or LCL will be considered as 'out of control' and likely caused by special events

# Control Chart: Xbar-R

Rational subgroups are samples that can be used to assess:

- Within group variation
- Between groups variation

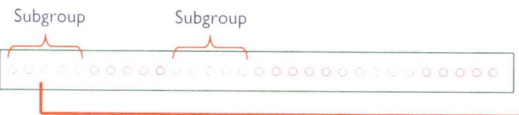

Subgroup        Subgroup

Control limits calculation:

### Xbar Chart

$$UCL = \overline{\overline{x}} + A_2 \overline{R}$$

$$CL = \overline{\overline{x}}$$

$$LCL = \overline{\overline{x}} - A_2 \overline{R}$$

### R Chart

$$UCL = D_4 \overline{R}$$

$$CL = \overline{R}$$

$$LCL = D_3 \overline{R}$$

where

| Sample size | Xbar - R Chart | | |
|---|---|---|---|
| | $A_2$ | $D_3$ | $D_4$ |
| 2 | 1,881 | 0 | 3,269 |
| 3 | 1,023 | 0 | 2,574 |
| 4 | 0,729 | 0 | 2,282 |
| 5 | 0,577 | 0 | 2,114 |
| 6 | 0,483 | 0 | 2,004 |
| 7 | 0,419 | 0,076 | 1,924 |
| 8 | 0,373 | 0,136 | 1,864 |
| 9 | 0,337 | 0,184 | 1,816 |
| 10 | 0,308 | 0,223 | 1,777 |

# Control Chart: Xbar-R

MINITAB:

<u>S</u>tat > <u>C</u>ontrol Charts > Variables Chart for <u>S</u>ubgroups > Xbar-R...

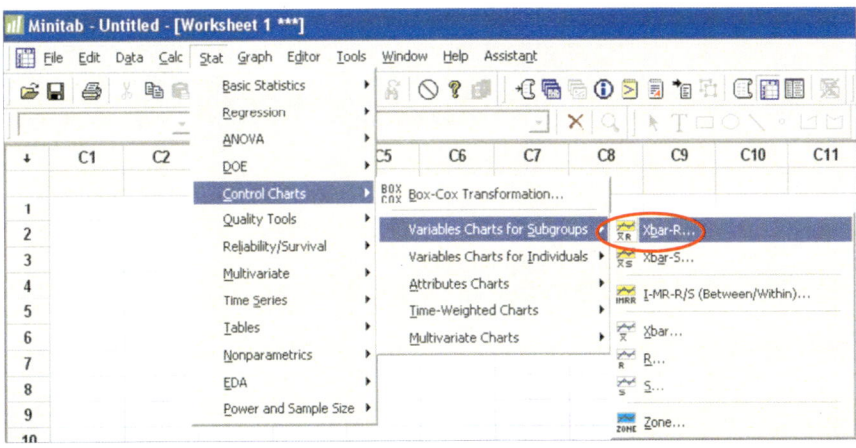

# Control Chart: Xbar-R

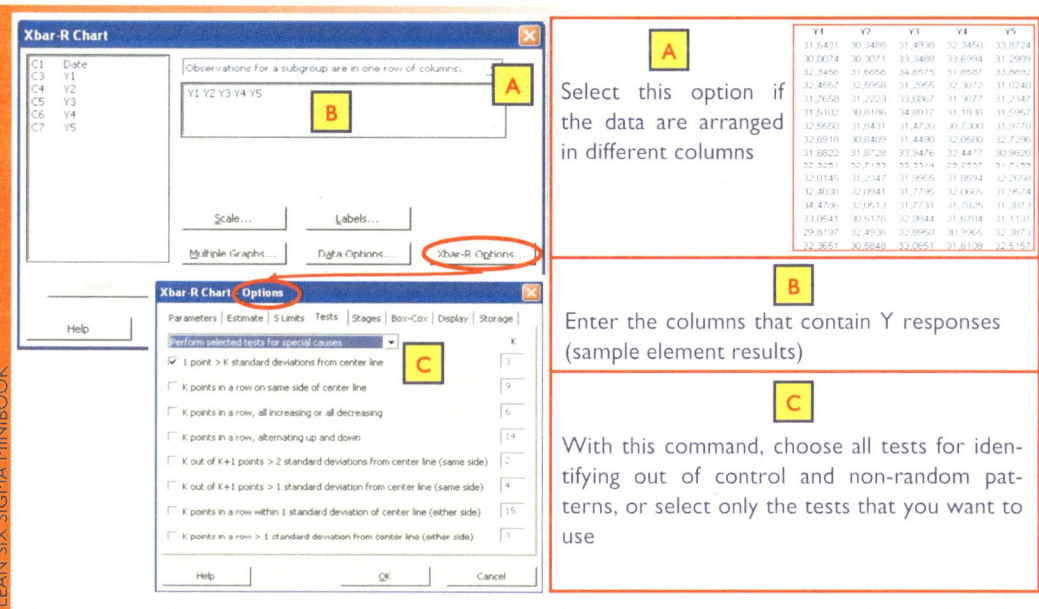

LEAN SIX SIGMA MINIBOOK

**A**

Select this option if the data are arranged in different columns

**B**

Enter the columns that contain Y responses (sample element results)

**C**

With this command, choose all tests for identifying out of control and non-random patterns, or select only the tests that you want to use

# Control Chart: Xbar-R

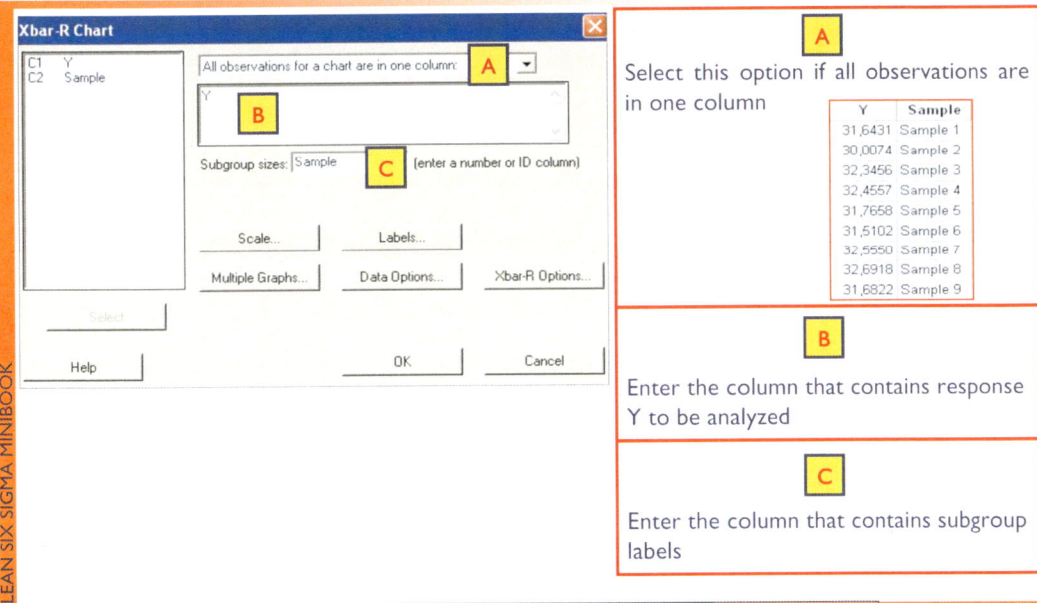

# Control Chart: Xbar-R

## MINITAB: Output

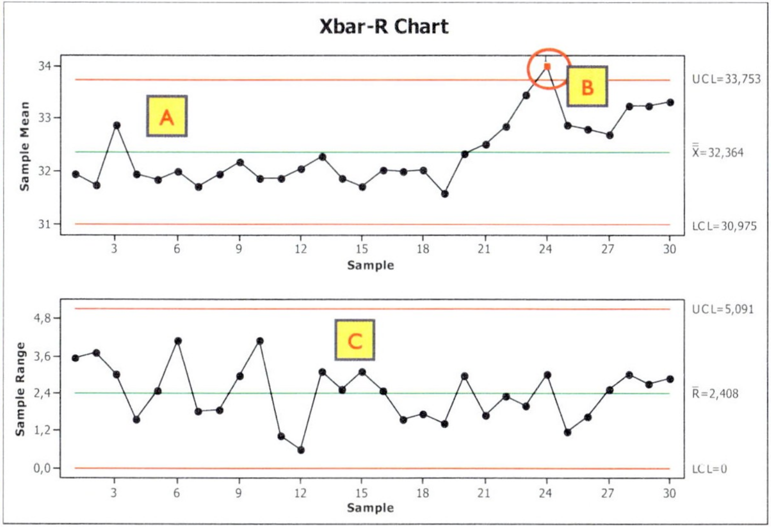

**Xbar Chart:** Each point in the chart represents a mean of a sample

Presence of a special cause

**R Chart:** Each point represents the range value in a sample

**Xbar-R Chart**

(Xbar Chart)
- Sample Mean axis: 31, 32, 33, 34
- UCL=33,753
- $\bar{\bar{X}}$=32,364
- LCL=30,975
- A
- B
- Sample axis: 3, 6, 9, 12, 15, 18, 21, 24, 27, 30

(R Chart)
- Sample Range axis: 0,0, 1,2, 2,4, 3,6, 4,8
- UCL=5,091
- $\bar{R}$=2,408
- LCL=0
- C
- Sample axis: 3, 6, 9, 12, 15, 18, 21, 24, 27, 30

# Control Chart: P Chart

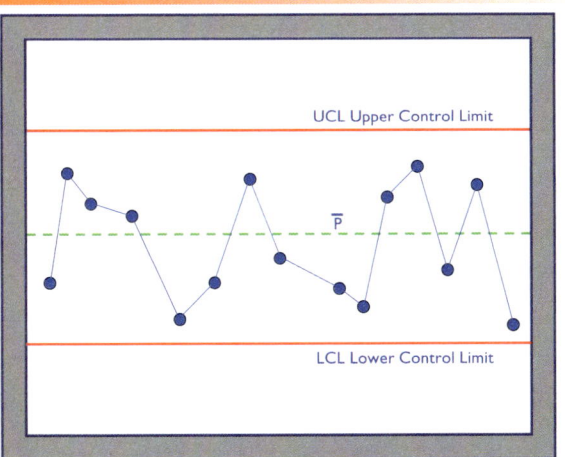

## When to use it:

- To monitor the state of inspection units, with 2 states, pass/failure, good/bad, etc. $\bar{p}$ is often the ratio of failure (discrete attribute variables)

$$UCL = \bar{p} + 3\sqrt{\frac{\bar{p}(1-\bar{p})}{n}}$$

$$CL = \bar{p}$$

$$LCL = \bar{p} - 3\sqrt{\frac{\bar{p}(1-\bar{p})}{n}}$$

Formula to calculate control limits

- Control limits are specified by the **Upper Control Limit** (UCL) and **Lower Control Limit** (LCL). They determine the range for natural variation due to random causes. Any variation beyond UCL or LCL will be considered as 'out of control' and likely caused by special events

# Control Chart: P Chart

MINITAB:

Stat > Control Charts > Attributes Chart > P...

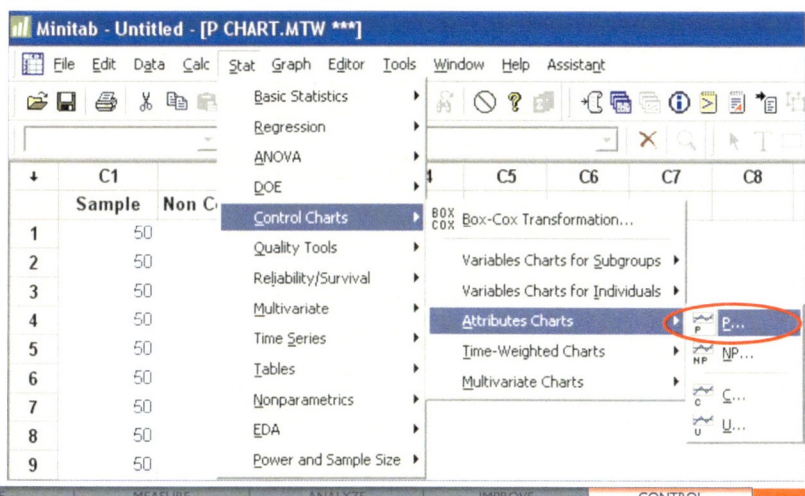

# Control Chart: P Chart

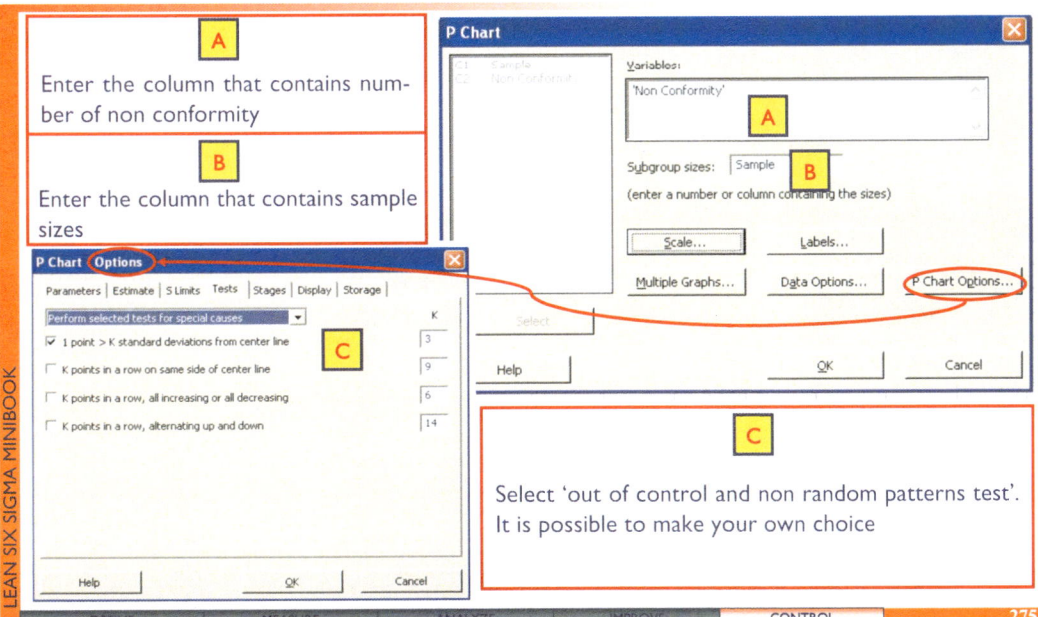

A
Enter the column that contains number of non conformity

B
Enter the column that contains sample sizes

**P Chart**

Variables:
'Non Conformity'

A

Subgroup sizes: Sample  B
(enter a number or column containing the sizes)

Scale... | Labels...

Multiple Graphs... | Data Options... | P Chart Options...

Select

Help | OK | Cancel

**P Chart Options**

Parameters | Estimate | S Limits | Tests | Stages | Display | Storage |

Perform selected tests for special causes

K

☑ 1 point > K standard deviations from center line    3

C

☐ K points in a row on same side of center line    9

☐ K points in a row, all increasing or all decreasing    6

☐ K points in a row, alternating up and down    14

Help | OK | Cancel

C
Select 'out of control and non random patterns test'.
It is possible to make your own choice

LEAN SIX SIGMA MINIBOOK

# Control Chart: P Chart

## MINITAB: Output

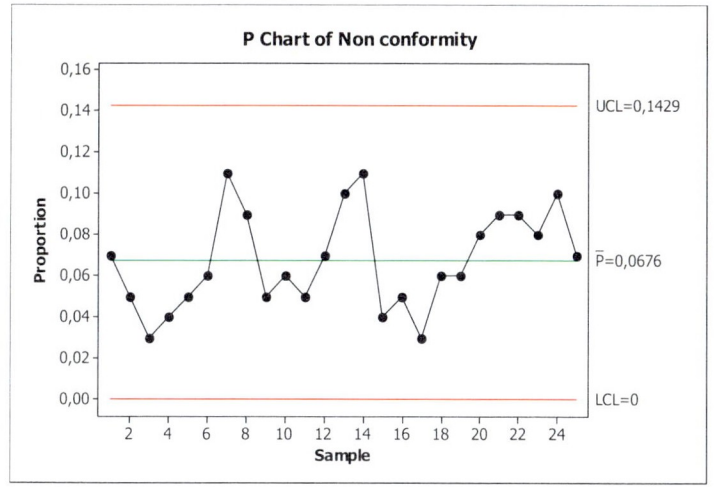

**P Chart of Non conformity**

UCL=0,1429

P̄=0,0676

LCL=0

| Sample | Non conformity |
|--------|----------------|
| 100 | 7 |
| 100 | 5 |
| 100 | 3 |
| 100 | 4 |
| 100 | 5 |
| 100 | 6 |
| 100 | 11 |
| 100 | 9 |
| 100 | 5 |
| 100 | 6 |
| 100 | 5 |
| 100 | 7 |
| 100 | 10 |
| 100 | 11 |
| 100 | 4 |
| 100 | 5 |
| 100 | 3 |
| 100 | 6 |
| 100 | 6 |
| 100 | 8 |
| 100 | 9 |
| 100 | 9 |
| 100 | 8 |
| 100 | 10 |
| 100 | 7 |

In this example, the process appears in the state of statistical control

LEAN SIX SIGMA MINIBOOK

# Control Chart & Minitab

Minitab Assistant helps you to choose the right Control Chart:

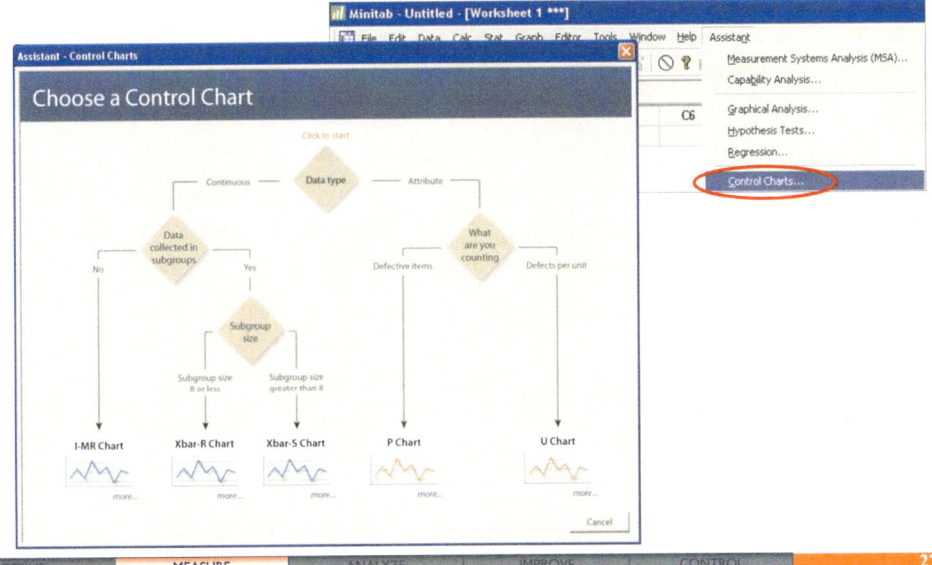

# Poka Yoke

Objective:

- Poka Yoke [G] (or *Mistake Proofing*) is one of the techniques that aims to reach the "Zero Defect Quality" through the usage of devices or procedures which allows detection of an error that could lead to waste (scraps, reworks, breakdowns etc.). Poka Yoke could be a design choice (Poka Yoke design) or a detection system on the process (Poka Yoke Process)

Poka Yoke basic concept:

- An error increases its economic impact if the time between when it happens and its discovery increases
- It is natural that people make mistakes but it is possible to prevent human errors becoming defects before they happen
- Don't try to do better next time: act now!

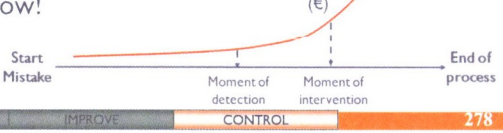

DEFINE | MEASURE | ANALYZE | IMPROVE | CONTROL

# Poka Yoke

## Why a mistake could happen:

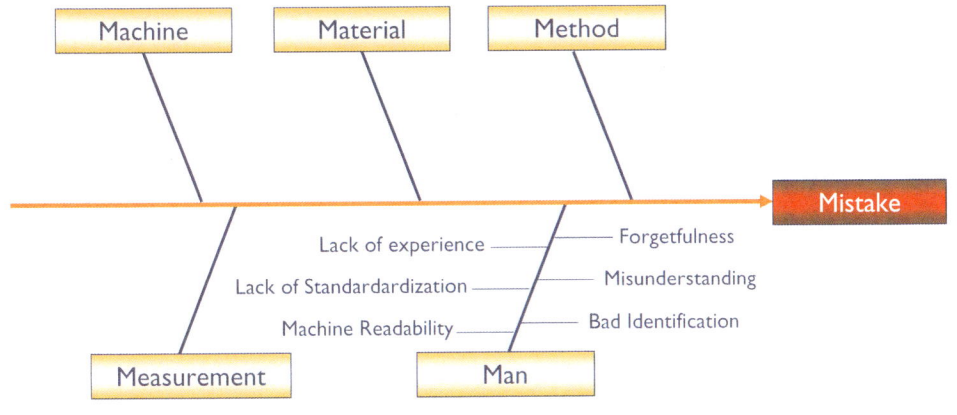

# Poka Yoke

Poka yoke strategy and examples:

| | | | |
|---|---|---|---|
| **MISTAKE** | Could happen... | Prevention area | |
| | Already happened... | Detection area | |

# Poka Yoke

## Types of Quality Inspections (Shingo):

1. **Judgment Inspection**:
   Discover defects after the facts, sort out bad products

   ### Disadvantages:
   - costly
   - difficult to eliminate all defects if human inspection is used
   - no feedback to process

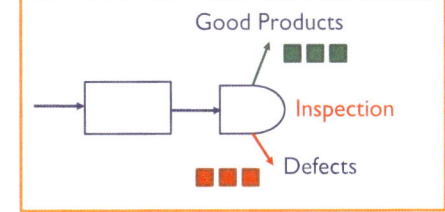

2. **Statistical Process Control (SPC)**:
   A sampling based inspection and process feedback

   ### Disadvantages of SPC (Shingo):
   - it is a sample inspection, so if the process is not capable, it cannot catch all the defects
   - the feedback is usually slow

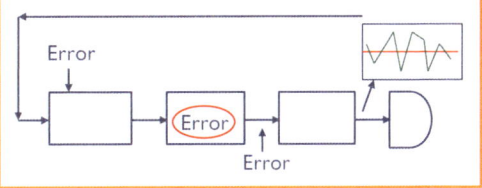

LEAN SIX SIGMA MINIBOOK

# Poka Yoke

## Desirable Types of Quality Inspections (Shingo):

### 1. Successive Check

Successive check is to let the immediate downstream process check output of the immediate upstream process

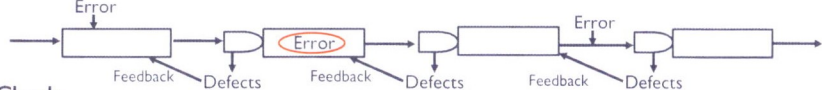

### 2. Self Check

Self check is to let current process check itself and trouble shoot the process immediately if defect occurs

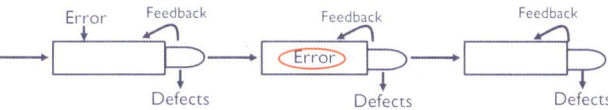

### 3. Source Inspection

Source inspection is based on the idea of discovering and correcting errors before errors become a defect

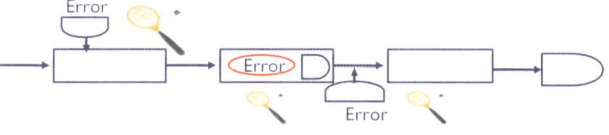

# Poka Yoke

## Basic Principles of Good Quality Inspections (Shingo):

1. Always 100% Inspection
2. Judgment about defects should be done objectively
3. Inspection should be done automatically at low cost
4. When a defect occurs, information should be immediately given as feedback and root causes immediately investigated
5. Discover hidden root causes one at a time, and eliminate them one by one
6. It is desirable to track the source of root cause therefore automatic inspection devices should be installed (Source Inspection)

## Poka Yoke System for Perfect Quality Inspection:

1. Poka Yoke device should be cheap, able to perform 100% inspection, and gives results instantly
2. Poka Yoke system can detect abnormalities by detection technology and/or process procedures

# Poka Yoke

## 3 Types of Regulatory Mechanisms for Poka Yoke:

### 1. Control:

When abnormalities occur, the Poka Yoke system will halt the operations

**Example:  Coffee Pot**

**Type of Inspection**:
Source Inspection

**Poka Yoke Detection Mechanism**:
Cheap Device

This automatic coffee pot is equipped with a "pause and serve" feature.  The lid of the pot presses up the button (shown above right), releasing a valve which allows the coffee to pour into the pot. If you remove the pot before brewing is finished, the valve closes, so coffee is not spilled

# Poka Yoke

## 3 Types of Regulatory Mechanisms for Poka Yoke:

### 2. Warning:
When abnormalities occur, the Poka Yoke system will send warning

### Example 1: Bathroom Poka Yoke

Some people may hang coat in a bathroom stall and then walk out without it. A stall door in a bathroom is designed so you cannot unlatch the door without dropping your coat unless you remove the coat first

**Type of Inspection**: Source Inspection
**Poka Yoke Detection Mechanism**: Smart layout design

### Example 2: Warning Color

In a hospital, the circular ring changes from purple to pink when water exceeds 40 degrees Celsius to avoid hurting patients

# Poka Yoke

## 3 Types of Regulatory Mechanisms for Poka Yoke:
### 3. Mistake proof:

When abnormalities occur, the Poka Yoke system will make mistakes impossible to happen

### Example 1: Electrical Plug

British 240v/50Hz electricity can injure people so electrical plugs are designed so that live electrical pins are never exposed:

1. the position and orientation of the pin are such that the plug can only fit one way in the socket

2. the pins are insulated near the plug body so that an electric shock is not possible via the exposed pin if the plug is not pushed all the way in but still making contact

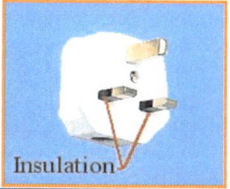

Insulation

### Example 2: Signatures Sheet

| AUTHORIZING SIGNATURES | |
|---|---|
| Operations Engineering Manager | Date |
| EM/EQA Supervisor | Date |
| Manufacturing Engineer | Date |
| Other (As Required) | Date |
| Operations Manager (MED. or HIGH RISK ONLY) | Date |
| EQA Manager (MED. or HIGH RISK ONLY) | Date |

Engineering change form requires different signatures depending on the nature of the change being considered. Sometimes engineers would get too many signatures, and sometimes not enough

| Required Authorizing Signatures | Release Audit update per SOP <or> Experiment Request | ECO Implementation | Part Renumbering per SOP (2-FP-51-000) | Non-Eco associated PCN RISK VERY LOW, LOW (1,2) | Non-ECO associated PCN RISK MODERATE, HIGH (3,4) | DATE |
|---|---|---|---|---|---|---|
| Originator | | | | | | |
| CCB Eng. Rep. | | | | | | |
| Ops. Supervisor* | | | | | | |
| QA Supervisor* SQE | | | | | | |
| Prod. Ops Manager | | | | | | |
| QA Manager | | | | | | |
| Manuf. Eng or Other ex. CTS (as required) | | | | | | |
| * Supervisor of area affected by the change | | | | | | |

The revised form identifies the nature of the change in the columns and indicates unnecessary signatures in gray. Creating forms that help the user fill them out correctly is a part of mistake-proofing

# Visual Management

## Objective:

- Visual Management[G] is a method that makes all processes in a company visual and tangible. Applying visual management will make the workplace well structured and processes will be clear to everybody at all levels (from top management to shopfloor). In other words "Make it visual"

## Why to use it:

- Visual Management has many advantages:
  - The detection of normal and abnormal operating conditions is easy and rapid
  - Time reduction
  - Space reduction
  - Costs reduction
  - … in general *waste reduction*

# Visual Management

How to implement Visual Management:

1. Decide what message is necessary to send or which mistakes you want to prevent from happening. During this step it is really important to identify communication at all levels

2. Design a simple visual "tool" to clearly communicate your message (colored lines, colored bins; visual dashboard; Andon; Kanban card, gage, checklist etc.). It is necessary to involve people

3. Test the visual impact on people and ask for feedback from those involved

4. Make all the adjustments or changes necessary to improve the communication effects

# Visual Management

## Visual Management examples:

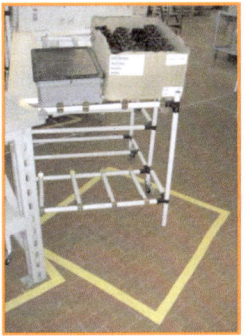

Not standard condition: the material rack is not in the designated place

Standard condition: the material rack is in its designated place

Files are organized with colored labels (each color represents a different year)

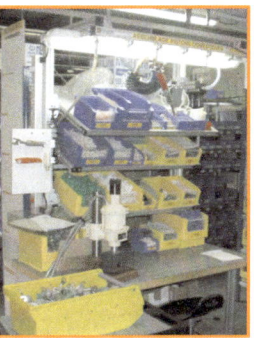

Blue bins indicate raw material. Yellow bins indicate WIP materials

# Visual Management

## Visual Management examples:

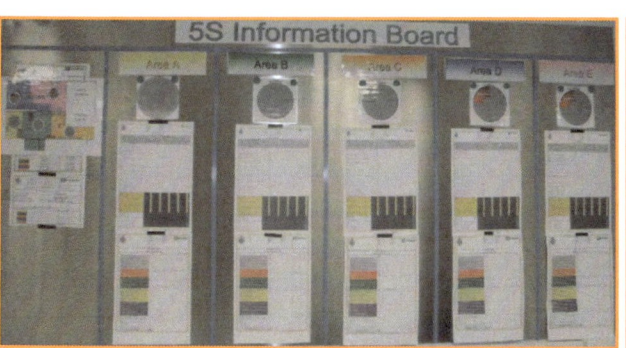

5S Dashboard: the information board allows employees at all levels to see the 5S status for each area in which 5S method is implemented

Red lines and labels indicate a place where scrap parts are segregated

# Visual Management

## Visual Management examples:

Strategy deployment activity board: Team department dashboard to assess the follow up phase of the annual strategy deployment cycle

# Visual Management

## Visual Management examples:

Manufacturing cell information board

"Andon (G)" for production control

# OPL - One Point Lesson

## Objective:

- The One Point Lesson (OPL[G]) is a method that allows a rapid and effective transfer of information from the leader of a group to its members. It must be written as simple as possible and should follow the 80-20 rule and use 80% pictures and 20% words (or less). The lesson should take from 5 to 10 minutes

## Types of OPL:

- **Basic information One Point Lesson**: used to transfer essential basic information, practical know-how and know-how of methods
- **Problem/Defect One Point Lesson**: used to teach how to prevent occurrence of a specific problem
- **Improvement/Kaizen**: describes the main characteristic and key measures in an improvement activity

# OPL - One Point Lesson

## One point lesson formats

# OPL - One Point Lesson

## Why is OPL important?

- To train people fast on a precise item
- To strengthen their understanding of the functions of machines, lines and processes in general (both manufacturing and transactional)
- To teach people about specific problems/defects in the process in order to eliminate their occurrence

## Who should use the OPL?

- Team leader or process expert

## Where should the OPL activities happen?

- At the "Gemba" (shopfloor, next to the machine, next to the computer in the case of software training, etc.)

# OPL - One Point Lesson

## One point lesson example

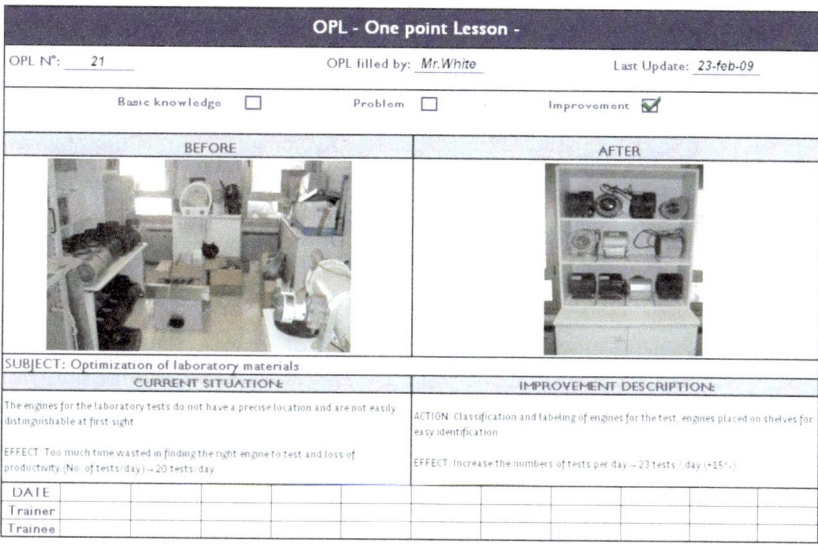

| OPL - One point Lesson - | | |
|---|---|---|
| OPL N°: 21 | OPL filled by: Mr.White | Last Update: 23-feb-09 |

| Basic knowledge ☐ | Problem ☐ | Improvement ☑ |
|---|---|---|

| BEFORE | AFTER |
|---|---|

SUBJECT: Optimization of laboratory materials

| CURRENT SITUATION: | IMPROVEMENT DESCRIPTION: |
|---|---|
| The engines for the laboratory tests do not have a precise location and are not easily distinguishable at first sight.<br><br>EFFECT: Too much time wasted in finding the right engine to test and loss of productivity (No. of tests/day) → 20 tests/day | ACTION: Classification and labeling of engines for the test, engines placed on shelves for easy identification<br><br>EFFECT: Increase the numbers of tests per day → 23 tests / day (+15%) |

| DATE | | | | | | | |
|---|---|---|---|---|---|---|---|
| Trainer | | | | | | | |
| Trainee | | | | | | | |

# OPL - One Point Lesson

## One point lesson example

| OPL - One Point Lesson - | | |
|---|---|---|

| OPL N°:  2 | OPL filled by: Mr.Green | Update: 31. Jul 08 |
|---|---|---|

| | Basic knowledge | ☑ | Problem | ☐ | Improvement | ☐ |
|---|---|---|---|---|---|---|

**SUBJECT:**

| | **DESCRIPTION:** |
|---|---|

Data Invio a Referente di Customer Care 14/07/2008  Data ultimo sollecito 14/07/2008  Num. solleciti 1  Data risposta del Customer Care 14/07/2008
Area responsabile Annunciatazione-Vendita/Direct  Responsabilità ☐ Sì ☑ No  Gravità reclamo 4  email Customer Care
Tipo reclamo finale Ritardo e o cattiva gestione della consegna  Contenuto risposta del DC  Ricarda Scorre su data  Crea Modulo N.C.
Contenuto finale Problematiche alla Consegna del veicolo  14/07/2008

*Procedure for the correct allocation of responsibility in a complaint*

**MAIN ACTIONS:**

*Select "YES" if the complaint responsibility is attributable to XYZ; select "NO" if the complaint is not of XYZ's responsibility.*

*ATTENTION: Assign the responsibility of the complaint only if after receiving the response from the department DC*

| DATE | | | | | | | | |
|---|---|---|---|---|---|---|---|---|
| Trainer | | | | | | | | |
| Trainee | | | | | | | | |

# Lean Six Sigma Checklist

Arcidiacono G., Calabrese C., Yang K.: Leading processes to lead companies: Lean Six Sigma.
DOI 10-1007/978-88-470-2492-2, © Springer-Verlag Italia 2012

# Define Checklist

Some steps to check *Define* phase completion:

| | |
|---|---|
| | Project selected with *Sponsor* |
| | Business Case identified |
| | Process identified |
| | Customer/s identification completed |
| | Team Leader choosen (Green Belt/Black Belt; internal/External) |
| | Team Member defined |
| | Project scope identified |
| | Project constraints identified and examined |
| | VOC (Voice of Customer) identified |
| | CTQs defined (SMART) |
| | SIPOC developed |
| | "As is" process mapping built and shared |
| | Benefit estimation (hard and soft savings) defined |
| | Waste properly identified |
| | Project Charter completed and shared with Sponsor |

 = Activity completed

# Measure Checklist

Some steps to check *Measure* phase completion:

| | |
|---|---|
| | Output operative definition identified |
| | Xs (potential input variables) to be measured identified |
| | Xs operative definition identified |
| | Setup of data collection plan |
| | Measurement system validated (Are the data reliable?) |
| | Data collection implemented and completed |
| | Basic statistics analysis & Trend analysis completed |
| | Process Capability |
| | Process performance identified (OEE; Takt time; Process Sigma; etc.) |
| | Measure phase shared with Sponsor |

# Analyze Checklist

Some steps to check *Analyze* phase completion:

| | |
|---|---|
| | **QUALITATIVE ANALYSIS** |
| | Potential root causes defined |
| | Stratification factors explored |
| | Process mapping analysis completed |
| | "Root causes" identified |
| | **QUANTITATIVE ANALYSIS** |
| | Significance of root causes validated, "Vital Few" identified (Regression, Hypothesis Testing) |
| | Analyze phase and global project review shared with Sponsor |

# Improve Checklist

Some steps to check *Improve* phase completion:

| | |
|---|---|
| | Solution generation (DOE, Creative Thinking; Lean applications) |
| | Solution evaluation completed |
| | Optimal solution identified and shared with Sponsor |
| | Lean application implemented (5S; Standard Work; Kanban; etc.) |
| | Pilot and check completed |
| | Pilot: potential risks identified |
| | Solution implementation validated |
| | Improve phase shared with Sponsor |

# Control Checklist

Some steps to check *Control* phase completion:

| | |
|---|---|
| ☐ | Standard procedure for CTQs monitoring implemented (Control plan) |
| ☐ | Standard procedures documented |
| ☐ | Visual management and error proofing systems set and implemented |
| ☐ | Potential improvement areas identified (Continuous Improvement) |
| ☐ | "Lesson Learned" prepared |
| ☐ | Final project review with Sponsor |
| ☐ | Project closure |
| ☐ | Summarized project documentation completed |
| ☐ | Extended project documentation completed |
| ☐ | Celebration |

# APPENDIX A: Process Sigma Table (I)

| Yield | DPMO | Sigma |
|---|---|---|
| 6,68072% | 933.193 | 0,00 |
| 7,35293% | 926.471 | 0,05 |
| 8,07567% | 919.243 | 0,10 |
| 8,85080% | 911.492 | 0,15 |
| 9,68005% | 903.200 | 0,20 |
| 10,56498% | 894.350 | 0,25 |
| 11,50697% | 884.930 | 0,30 |
| 12,50719% | 874.928 | 0,35 |
| 13,56661% | 864.334 | 0,40 |
| 14,68591% | 853.141 | 0,45 |
| 15,86553% | 841.345 | 0,50 |
| 17,10561% | 828.944 | 0,55 |
| 18,40601% | 815.940 | 0,60 |
| 19,76625% | 802.337 | 0,65 |
| 21,18554% | 788.145 | 0,70 |
| 22,66274% | 773.373 | 0,75 |
| 24,19637% | 758.036 | 0,80 |
| 25,78461% | 742.154 | 0,85 |
| 27,42531% | 725.747 | 0,90 |
| 29,11597% | 708.840 | 0,95 |

| Yield | DPMO | Sigma |
|---|---|---|
| 30,85375% | 691.462 | 1,00 |
| 32,63552% | 673.645 | 1,05 |
| 34,45783% | 655.422 | 1,10 |
| 36,31693% | 636.831 | 1,15 |
| 38,20886% | 617.911 | 1,20 |
| 40,12937% | 598.706 | 1,25 |
| 42,07403% | 579.260 | 1,30 |
| 44,03823% | 559.618 | 1,35 |
| 46,01722% | 539.828 | 1,40 |
| 48,00612% | 519.939 | 1,45 |
| 50,00000% | 500.000 | 1,50 |
| 51,99388% | 480.061 | 1,55 |
| 53,98278% | 460.172 | 1,60 |
| 55,96177% | 440.382 | 1,65 |
| 57,92597% | 420.740 | 1,70 |
| 59,87063% | 401.294 | 1,75 |
| 61,79114% | 382.089 | 1,80 |
| 63,68307% | 363.169 | 1,85 |
| 65,54217% | 344.578 | 1,90 |
| 67,36448% | 326.355 | 1,95 |

| Yield | DPMO | Sigma |
|---|---|---|
| 69,14625% | 308.538 | 2,00 |
| 70,88403% | 291.160 | 2,05 |
| 72,57469% | 274.253 | 2,10 |
| 74,21539% | 257.846 | 2,15 |
| 75,80363% | 241.964 | 2,20 |
| 77,33726% | 226.627 | 2,25 |
| 78,81446% | 211.855 | 2,30 |
| 80,23375% | 197.663 | 2,35 |
| 81,59399% | 184.060 | 2,40 |
| 82,89439% | 171.056 | 2,45 |
| 84,13447% | 158.655 | 2,50 |
| 85,31409% | 146.859 | 2,55 |
| 86,43339% | 135.666 | 2,60 |
| 87,49281% | 125.072 | 2,65 |
| 88,49303% | 115.070 | 2,70 |
| 89,43502% | 105.650 | 2,75 |
| 90,31995% | 96.800 | 2,80 |
| 91,14920% | 88.508 | 2,85 |
| 91,92433% | 80.757 | 2,90 |
| 92,64707% | 73.529 | 2,95 |

# APPENDIX A: Process Sigma Table (II)

| Yield | DPMO | Sigma |
|---|---|---|
| 93,31928% | 66.807 | 3,00 |
| 93,94292% | 60.571 | 3,05 |
| 94,52007% | 54.799 | 3,10 |
| 95,05285% | 49.471 | 3,15 |
| 95,54345% | 44.565 | 3,20 |
| 95,99408% | 40.059 | 3,25 |
| 96,40697% | 35.930 | 3,30 |
| 96,78432% | 32.157 | 3,35 |
| 97,12834% | 28.717 | 3,40 |
| 97,44119% | 25.588 | 3,45 |
| 97,72499% | 22.750 | 3,50 |
| 97,98178% | 20.182 | 3,55 |
| 98,21356% | 17.864 | 3,60 |
| 98,42224% | 15.778 | 3,65 |
| 98,60966% | 13.903 | 3,70 |
| 98,77755% | 12.224 | 3,75 |
| 98,92759% | 10.724 | 3,80 |
| 99,06133% | 9.387 | 3,85 |
| 99,18025% | 8.198 | 3,90 |
| 99,28572% | 7.143 | 3,95 |

| Yield | DPMO | Sigma |
|---|---|---|
| 99,37903% | 6.209,7 | 4,00 |
| 99,46139% | 5.386,1 | 4,05 |
| 99,53388% | 4.661,2 | 4,10 |
| 99,59754% | 4.024,6 | 4,15 |
| 99,65330% | 3.467,0 | 4,20 |
| 99,70202% | 2.979,8 | 4,25 |
| 99,74449% | 2.555,1 | 4,30 |
| 99,78140% | 2.186,0 | 4,35 |
| 99,81342% | 1.865,8 | 4,40 |
| 99,84111% | 1.588,9 | 4,45 |
| 99,86501% | 1.349,9 | 4,50 |
| 99,88558% | 1.144,2 | 4,55 |
| 99,90324% | 967,6 | 4,60 |
| 99,91836% | 816,4 | 4,65 |
| 99,93129% | 687,1 | 4,70 |
| 99,94230% | 577,0 | 4,75 |
| 99,95166% | 483,4 | 4,80 |
| 99,95959% | 404,1 | 4,85 |
| 99,96631% | 336,9 | 4,90 |
| 99,97197% | 280,3 | 4,95 |

| Yield | DPMO | Sigma |
|---|---|---|
| 99,97674% | 232,6 | 5,00 |
| 99,98074% | 192,6 | 5,05 |
| 99,98409% | 159,1 | 5,10 |
| 99,98689% | 131,1 | 5,15 |
| 99,98922% | 107,8 | 5,20 |
| 99,99116% | 88,4 | 5,25 |
| 99,99277% | 72,3 | 5,30 |
| 99,99409% | 59,1 | 5,35 |
| 99,99519% | 48,1 | 5,40 |
| 99,99609% | 39,1 | 5,45 |
| 99,99683% | 31,7 | 5,50 |
| 99,99744% | 25,6 | 5,55 |
| 99,99793% | 20,7 | 5,60 |
| 99,99834% | 16,6 | 5,65 |
| 99,99867% | 13,3 | 5,70 |
| 99,99893% | 10,7 | 5,75 |
| 99,99915% | 8,5 | 5,80 |
| 99,99932% | 6,8 | 5,85 |
| 99,99946% | 5,4 | 5,90 |
| 99,99957% | 4,3 | 5,95 |
| 99,99966% | 3,4 | 6,00 |

# APPENDIX B: Kinds of variables

This summary could be useful for the correct selection of indicators during the implementation of a Lean Six Sigma project

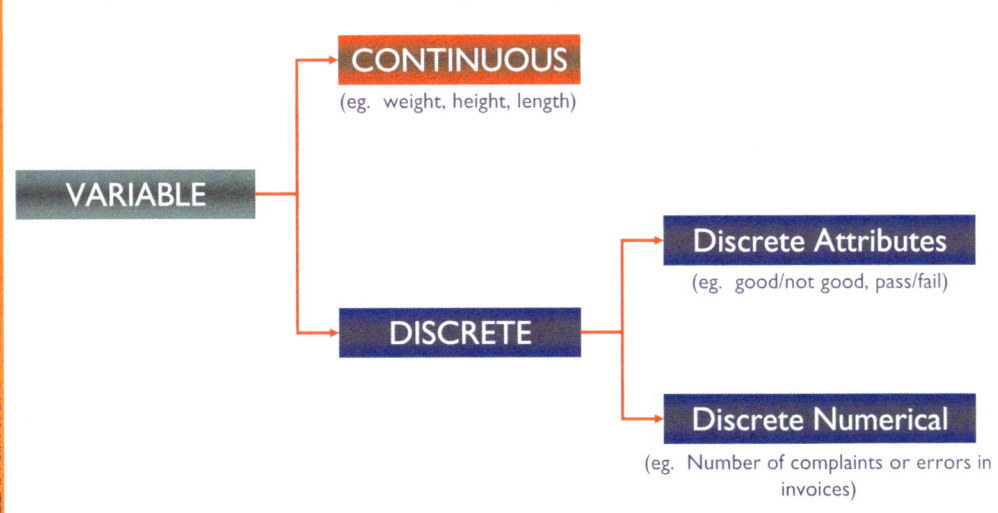

**CONTINUOUS**
(eg. weight, height, length)

**VARIABLE**

**DISCRETE**

**Discrete Attributes**
(eg. good/not good, pass/fail)

**Discrete Numerical**
(eg. Number of complaints or errors in invoices)

# APPENDIX C: Kaizen Leader Standard Form

## Waste Walk Analysis format example

| "Waste Walk Format" | | | | |
|---|---|---|---|---|
| Company: | | Product/Practice followed: | | |
| Observer: | | Date: | | |
| **Waste category** | **Process Step** | **Waste Comment** | **Waste estimation** | **Possible idea/solution** |
| | | | | |
| | | | | |
| | | | | |
| | | | | |
| | | | | |
| | | | | |

5S format example

| | 5S Program Areas | | | |
|---|---|---|---|---|
| | | PLANIMETRY... | | |
| # | AREA | AREA RESPONSIBLE | DEPUTY-RESPONSIBLE | TEAM |
| 1 | Pink | | | |
| 2 | Heavenly | | | |
| 3 | Green | | | |
| 4 | Blue | | | |
| 5 | Orange | | | |
| 6 | Yellow | | | |
| 7 | Grey | | | |
| 8 | Brown | | | |
| 9 | White | | | |

# APPENDIX C: Kaizen Leader Standard Form

## 5S format example

| Audit team and calendar | | | | | | | | | | | | | |
|---|---|---|---|---|---|---|---|---|---|---|---|---|---|

**Steering team members:**

Plant Manager: ........................................
Project Sponsor: ........................................
Lean coordinator: ........................................
Functional manager: ........................................
Department manager: ........................................

| # | AREA | G | F | M | A | M | J | L | A | S | O | N | D | Notes |
|---|------|---|---|---|---|---|---|---|---|---|---|---|---|-------|
| 1 | Pink | | | | | | | | | | | | | For each area, the audit teams should be built as follows: |
| 2 | Heavenly | | | | | | | | | | | | | |
| 3 | Black | | | | | | | | | | | | | - Area responsible |
| 4 | Blue | | | | | | | | | | | | | - External area responsible |
| 5 | Orange | | | | | | | | | | | | | - Two members of the steering team |
| 6 | Yellow | | | | | | | | | | | | | |
| 7 | Grey | | | | | | | | | | | | | |
| 8 | Brown | | | | | | | | | | | | | |
| 9 | White | | | | | | | | | | | | | |

# APPENDIX C: Kaizen Leader Standard Form

## 5S format example

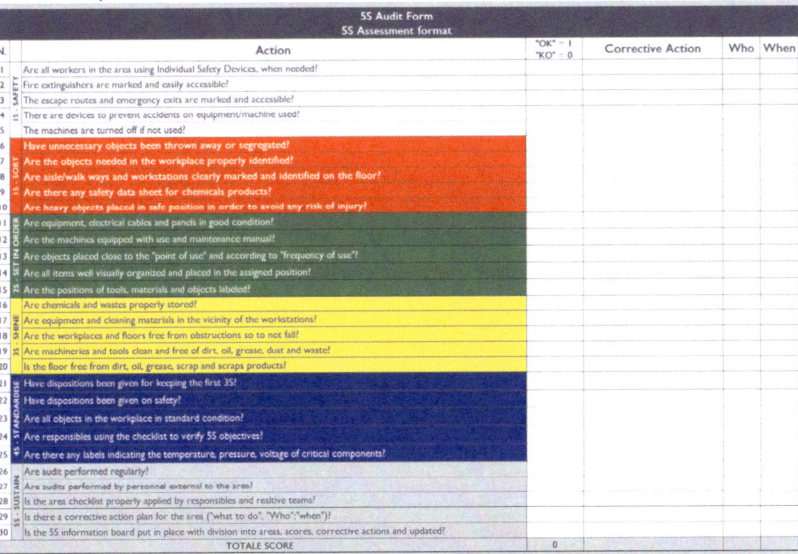

**5S Audit Form**
**5S Assessment format**

| N. | Action | "OK" = 1 "KO" = 0 | Corrective Action | Who | When |
|---|---|---|---|---|---|
| 1 | Are all workers in the area using Individual Safety Devices, when needed? | | | | |
| 2 | Fire extinguishers are marked and easily accessible? | | | | |
| 3 | The escape routes and emergency exits are marked and accessible? | | | | |
| 4 | There are devices to prevent accidents on equipment/machine used? | | | | |
| 5 | The machines are turned off if not used? | | | | |
| 6 | Have unnecessary objects been thrown away or segregated? | | | | |
| 7 | Are the objects needed in the workplace properly identified? | | | | |
| 8 | Are aisle/walk ways and workstations clearly marked and identified on the floor? | | | | |
| 9 | Are there any safety data sheet for chemicals products? | | | | |
| 10 | Are heavy objects placed in safe position in order to avoid any risk of injury? | | | | |
| 11 | Are equipment, electrical cables and panels in good condition? | | | | |
| 12 | Are the machines equipped with use and maintenance manual? | | | | |
| 13 | Are objects placed close to the "point of use" and according to "frequency of use"? | | | | |
| 14 | Are all items well visually organized and placed in the assigned position? | | | | |
| 15 | Are the positions of tools, materials and objects labeled? | | | | |
| 16 | Are chemicals and wastes properly stored? | | | | |
| 17 | Are equipment and cleaning materials in the vicinity of the workstations? | | | | |
| 18 | Are the workplaces and floors free from obstructions so to not fall? | | | | |
| 19 | Are machineries and tools clean and free of dirt, oil, grease, dust and waste? | | | | |
| 20 | Is the floor free from dirt, oil, grease, scrap and scraps products? | | | | |
| 21 | Have dispositions been given for keeping the first 3S? | | | | |
| 22 | Have dispositions been given on safety? | | | | |
| 23 | Are all objects in the workplace in standard condition? | | | | |
| 24 | Are responsibles using the checklist to verify 5S objectives? | | | | |
| 25 | Are there any labels indicating the temperature, pressure, voltage of critical components? | | | | |
| 26 | Are audit performed regularly? | | | | |
| 27 | Are audits performed by personnel external to the area? | | | | |
| 28 | Is the area checklist properly applied by responsibles and relative teams? | | | | |
| 29 | Is there a corrective action plan for the area ("what to do", "Who","when")? | | | | |
| 30 | Is the 5S information board put in place with division into areas, scores, corrective actions and updated? | | | | |
| | **TOTALE SCORE** | **0** | | | |

# APPENDIX C: Kaizen Leader Standard Form

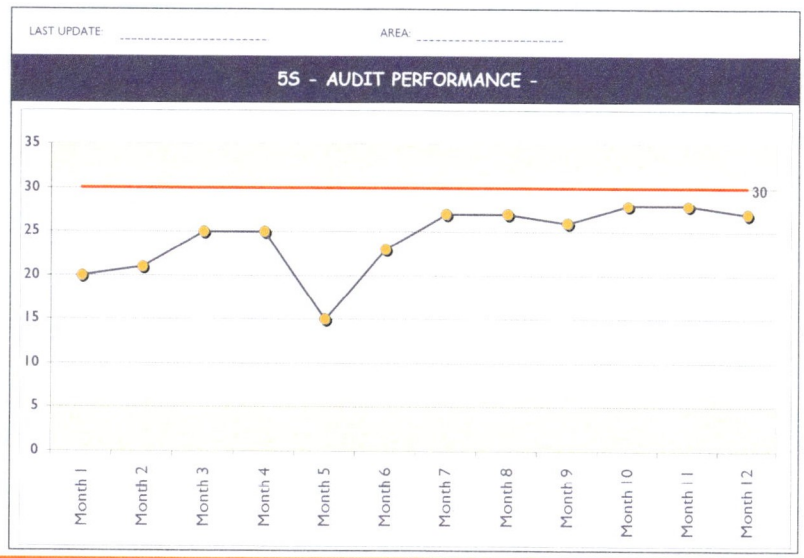

5S format example

LAST UPDATE: .......................... AREA: ..........................

**5S - AUDIT PERFORMANCE -**

# APPENDIX C: Kaizen Leader Standard Form

## Standard Work: Cycle Time Observation Form

| Observation Date | | Kaizen Leader | |
|---|---|---|---|
| Company | | Kaizen Team | |
| Kaizen | | | |
| Observation Time | | | |
| Process | | | |

| | | **Cycle Time Observation Form** | | | | | | | | | | | | |
|---|---|---|---|---|---|---|---|---|---|---|---|---|---|---|
| N. | Task | Cycle 1 | Cycle 2 | Cycle 3 | Cycle 4 | Cycle 5 | Cycle 6 | Cycle 7 | Cycle 8 | Cycle 9 | Cycle 10 | Mean | Lowest repeatable | Standard time |
| 1 | | | | | | | | | | | | | | |
| 2 | | | | | | | | | | | | | | |
| 3 | | | | | | | | | | | | | | |
| 4 | | | | | | | | | | | | | | |
| 5 | | | | | | | | | | | | | | |
| 6 | | | | | | | | | | | | | | |
| 7 | | | | | | | | | | | | | | |
| 8 | | | | | | | | | | | | | | |
| 9 | | | | | | | | | | | | | | |
| Cycle time (1 Cycle) | | 0 | 0 | 0 | 0 | 0 | 0 | 0 | 0 | 0 | 0 | | | 0 |

# APPENDIX C: Kaizen Leader Standard Form

## Standard Work: Process Capacity Form

| Observation Date | | Opearting time per shift | |
|---|---|---|---|
| Company | | Shift No | |
| Part No | | Daily demand | |
| Process | | Supervisor | |

| | | | | | | | | | | |
|---|---|---|---|---|---|---|---|---|---|---|
| **Process Capacity Form** | | | | | | | | | | |
| Step # | Process Description | Machine | Base Time | | Tool Change | | | Total Time (sec.) | Processing Capacity | Remarks |
| | | | Manual Time (sec.) | Machine Time (sec.) | # Pieces /Change | Change Time | Tool Change Time (sec.) | | | |
| 1 | | | | | | | | | | |
| 2 | | | | | | | | | | |
| 3 | | | | | | | | | | |
| | | | | | | | | | | |
| | | | | | | | | | | |
| | | | | | | | | | | |
| | | | | | | | | | | |
| | | Total: | 0 | | Maximum daily production: | | | | 0 | |

# APPENDIX C: Kaizen Leader Standard Form

## Standard Work: Standard Work Combination Form

| Observation Date | | Opearting time per shift | | – – – = Machine |
| Company | | Daily demand | | ▬▬ = Manual |
| Part No | | Takt Time (ses) | | ∿∿ = Walking |
| Process | | Supervisor | | ▬▬ = Takt Time |

### Standard Work Combination Form

| Step # | Task/Activity Description | Time | | | Operation Working Time (secs) |
|---|---|---|---|---|---|
| | | Man. | Auto | Walk | 5 10 15 20 25 30 35 40 45 50 55 60 65 70 75 80 85 90 95 100 105 |
| 1 | | | | | |
| 2 | | | | | |
| 3 | | | | | |
| 4 | | | | | |
| 5 | | | | | |
| 6 | | | | | |
| 7 | | | | | |
| 8 | | | | | |
| 9 | | | | | |
| 10 | | | | | |
| 11 | | | | | |
| 12 | | | | | |
| 13 | | | | | |

# APPENDIX C: Kaizen Leader Standard Form

## OPL - Basic information

| OPL - One point Lesson - | | |
|---|---|---|
| OPL N°: _____ | OPL filled by: _____ | Update: _____ |
| Basic knowledge ☑ | Problem ☐ | Improvement ☐ |

**SUBJECT:**

| | DESCRIPTION: |
|---|---|
| | |

**MAIN ACTIONS:**

| DATE | | | | | | | | |
|---|---|---|---|---|---|---|---|---|
| Trainer | | | | | | | | |
| Trainee | | | | | | | | |

# APPENDIX C: Kaizen Leader Standard Form

## OPL - Problem/Defect

| OPL - One point Lesson - | | |
|---|---|---|
| OPL N°: _____ | OPL filled by: _____ | Last Update: _____ |
| Basic knowledge ☐ | Problem ☑ | Improvement ☐ |

| BEFORE | AFTER |
|---|---|
| | |

SUBJECT:

PHENOMENON:

| CAUSES DESCRIPTION | MAIN ACTION |
|---|---|
| | |

| DATE | | | | | | | | |
|---|---|---|---|---|---|---|---|---|
| Trainer | | | | | | | | |
| Trainee | | | | | | | | |

# APPENDIX C: Kaizen Leader Standard Form

## OPL - Improvement/Kaizen

| OPL - One point Lesson - | | |
|---|---|---|
| OPL N°: _____ | OPL filled by: _____ | Last Update: _____ |

Basic knowledge ☐  Problem ☐  Improvement ☑

| BEFORE | AFTER |
|---|---|
|  |  |

SUBJECT:

| CURRENT SITUATION: | IMPROVEMENT DESCRIPTION: |
|---|---|
|  |  |

| DATE | | | | | | | | |
|---|---|---|---|---|---|---|---|---|
| Trainer | | | | | | | | |
| Trainee | | | | | | | | |

# APPENDIX C: Kaizen Leader Standard Form

## Kaizen Newspaper example:

| | | |
|---|---|---|
| **Kaizen event date** | 20 March 2009 | |
| **Company** | ABC Company | |
| **Place** | XXX | |
| **Kaizen** | Changeover reduction | |
| **Kaizen Leader** | Mr. Green | |
| **Kaizen Team** | Mr. Green; Mr. Yellow; Mrs. Orange; Mr. Blue; Mr. White | |
| **Last Update** | 30 March 2009 | |
| **% Kaizen** | 48% | |

### Kaizen Newspaper

| N. | Action | Who | Start date | Expected end date | End date | Priority | Status | Notes |
|---|---|---|---|---|---|---|---|---|
| 1 | Identify which families are more products on production line | Mr. Green | March 20 | March 23 | March 23 | A | 100% | |
| 2 | Determine which materials should be at each workstation | Mr. Yellow | March 20 | March 27 | | A | 56% | |
| 3 | Eliminate all the things not necessary | Mr. White | March 20 | March 21 | March 22 | A | 100% | |
| 4 | Identify which kind of bin should be used to manage the different part | Mr. Yellow | March 20 | April 4 | | A | 70% | Purchased n°30 bins model B; n°86 bins for small parts |
| 5 | Create workstation Before/After | Mr. Yellow | March 20 | April 4 | | A | 95% | |
| 6 | Label on the floor all the workstations | Team | March 20 | April 15 | | B | 0% | |
| 7 | Optimal layout study | Mr.Blue | March 20 | April 15 | | C | 0% | |

# APPENDIX C: Kaizen Leader Standard Form

Kaizen performance:

| | | | |
|---|---|---|---|
| | Kaizen event date | 25 May 2009 | |
| | Company | ABC Company | |
| | Place | XXX | |
| | Kaizen | Changeover reduction | |
| | Kaizen Leader | Mr. Green | |
| | Kaizen Team | Mr. Green; Mr. Yellow; Mrs. Orange; Mr. Blue; Mr. White | |
| | Last Update | 30 March 2009 | |
| | % Kaizen | 95% | |

### Kaizen Performance

| N. | Measurement | UoM | Before Kaizen | Kaizen Goal | After Kaizen | After 2 Weeks | After 4 Weeks | After 6 Weeks | After 8 Weeks |
|---|---|---|---|---|---|---|---|---|---|
| 1 | Productivity | Pcs/Hours | | | | | | | |
| 2 | Space reduction | m² | | | | | | | |
| 3 | Walking distance | m | | | | | | | |
| 4 | Batch size | Pcs | | | | | | | |
| 5 | Set up time | Min | 79' | 30' | 40' | 25' | 25' | 22' | 27' |
| 6 | Cycle time | Min/secs | | | | | | | |
| 7 | OEE | % | | | | | | | |
| 8 | Savings | (€) | | | | | | | |
| 9 | Lead Time (gg) | | | | | | | | |
| 10 | Inventory turn | | | | | | | | |
| 11 | 5S Score | | | | | | | | |

# Index

LEAN SIX SIGMA MINIBOOK

# Index

# Index

# Glossary

### A

**Andon**:  Andon is any visual indicator signaling the current status of a step in the production/process system. It alerts team leaders or supervisors in case of existing or emerging production/process problems.

### B

**Brainstorming**: A group based creativity technology that is designed to generate and select ideas for problems solving.

**BVA, Business Value Added activity**: Activity that does not add any value to the product/service but is necessary from a business operations' point of view.

### C

**Cell:** It is a workplace in which equipment, people, machinery, materials and methods are arranged in order to have continuous production flow.

**Confidence Interval (CI):** is the interval which, with a likely probability, contains the mean (or proportion, median, standard deviation) of the population, where the sample comes from.

**Common Cause:** The cause, random in nature and not related to any special event, is behind natural inherent variability displayed in processes.

# Glossary

**COPQ, Cost Of Poor Quality**: COPQ are the costs related to poor performance of manufacturing or transactional processes.

**CTQ, Critical To Quality**: The key measurable features of a product or process whose performance standards or specification limits must be met in order to satisfy the customer.

**Customer:** The client, internal or external, is the recipient of a process / product / service.

**Customer Satisfaction**: is a measure of how products and services supplied by a company meet customer expectation. Customer expectation should be objectively and accurately measured by collecting and analyzing "Voice Of the Customer" (VOC). It is the starting point for identifying improvements.

## D

**DMAIC**: Stands for 5 phases of Lean Six Sigma methodology: Define, Measure, Analyze, Improve, Control.

**DOE, Design Of Experiment:** DOE is a methodology that builds, through well-planned experiments and analysis of the experimental results, the analytical model relating to the cause-effect relationship between input and output variables.

**DPMO, Defects Per Million of Opportunity:** DPMO is a performance indicator calculated as a ratio of number of defects divided by maximum number of potential defects in a batch of units inspected.

# Glossary

## F

**FMEA, Failure Modes and Effects Analysis**: FMEA is a tool that can be used to identify a detailed list of failure modes of a product or process and their corresponding causes and then rate them with a severity level, a likelihood of occurrence and detection in order to manage the system risk.

## H

**Heijunka:** is one of the elements of Just in Time and it is the process of smoothing the type and quantity of production over a fixed period of time.

## J

**Jidoka:** This term means "automation with human intelligence". It means that an automated process is sufficiently "aware" of itself so that it will detect process malfunctions or product defects, stop itself and alert the operator.

## K

**Kaizen:** means "to become good through change". A Kaizen event is a focused effort for make an improvement activity.

**Kanban:** It is a method used in many applications in various processes. It is primarily used as an instruction mechanism that controls the production, movement of goods, material, or parts, or jobs.

# Glossary

## L

**LCL, Lower Control Limit:** Represents the lower limit of a stable distribution for the variability of a process (VOP).

**Lead Time:** is the time between the placing of an order and the receipt of goods/services ordered (it is also possible to speak about Production Lead Time, Delivery Lead Time etc.).

**Lean:** is the methodology that aims to identify and eliminate wastes in order to maximize speed and flexibility of business processes so we can deliver <u>what</u> is needed, <u>when</u> needed and in the <u>quantity</u> needed by the Customer.

**LSL, Lower Specification Limit:** Represents the lower limit of a tolerance region that is acceptable by the customer.

## N

**NVA, Non Value Added activity:** Activity that does not add any value to the product/service.

## O

**OEE, Overall Equipment Effectiveness:** is a powerful method to monitor and improve the efficiency of manufacturing and transactional processes.

# Glossary

**OPL, One Point Lesson:** is a method that allows a rapid and effective transfer of information from the leader of a group to its members.

**Outlier:** An observation that is numerically distant from the rest of the data.

P

**Poka Yoke:** It is one of the techniques that aims to reach the "Zero Defect Quality" through the usage of devices or procedures, which allows detection of an error that could lead to waste.

**Process Capability Analysis:** Also called Capability Analysis, is a performance index used to measure the ability of a process (VOP) to meet the specification limits defined by customers (VOC).

**Process Owner:** is the owner of the process, usually the head of a department or office in which the Lean Six Sigma project is implemented.

**Process Sigma:** Process Sigma is a performance metric that is based on comparing specification length with the standard deviation (Sigma) of process. This performance index is related to defective rates.

**Project Charter:** is a document which contains key information on implementing a Lean Six Sigma project.

**P-Value:** is a measure of how much evidence we have against the importance of a factor. The smaller the P-Value, the stronger the evidence. A P-Value of < 0.05 is an indication of statistically significant evidence.

# Glossary

R

**Rational subgroups:** The rational subgroups are samples chosen in a way that maximize the variability between samples when there are special causes present and the variability within the sample is minimized.

**Reorder Point (ROP):** is the inventory level of an item which signals the need for placement of a replenishment order, taking into account the consumption of the item during order Lead Time and the quantity required for safety stock.

**Residual:** Difference between actual value of data and predicted value from mathematical models (derived by Regression or Design Of Experiments).

S

**Savings:** Economic or strategic benefits resulting from improvement/project activities.

**SIPOC:** High level process mapping to describe any kind of process (Supplier, Input, Process, Output, Customer).

**Standard Work:** It is the most effective combination of manpower, materials and machinery to produce something in the time, quality and quantity required by customer.

# Glossary

**Six Sigma**: A well structured and disciplined operating strategy (structured according to the DMAIC phases), to measure, analyze and improve the performance in terms of operational excellence. The Six Sigma methodology is sufficiently flexible and adaptable to different business contexts.

**SMED**: Single Minute Exchange of Die is a method that aims to reduce the changeover time of equipment, machine or a production/service process in general.

**Special Cause**: The cause that is often associated with a special event and the result of a special cause often lets the process form a trend, seasonality or other non random patterns.

## T

**Takt Time**: It represents the rhythm of production/delivery that a process (workstation, Cell, etc.) must respect to satisfy customer demand.

## U

**UCL, Upper Control Limit**: Represents the upper limit of a stable distribution for the variability of a process (VOP).

**USL, Upper Specification Limit**: Represents the upper limit of a tolerance region that is acceptable by the customer.

# Glossary

## V

**VA, Value Added activity:** An activity that increases the value of the product/service from the customer's point of view. It is something that customers are willing to pay for.

**Visual Management:** It is a method that makes all processes within a company visual and tangible.

**VOC, Voice Of The Customer:** The "Voice of the Customer" is how the customer perceives the product/process/service in comparison with their desires.

**VOP, Voice Of Process:** The "Voice of the Process" is what the process/product/service is able to deliver.

**VSM, Value Stream Mapping:** It is a diagram of every step involved in the material and information flow necessary to bring the product/service from the order to delivery phase.

## W

**Waste:** It is the use of resources (time, material, labor, etc.) for doing something that the customers are not willing to pay for and, therefore, does not add value to the product or service provided.

# References

- Arcidiacono G. (2006) *Keys to success for Six Sigma*, Proceedings of ICAD2006, Fourth International Conference on Axiomatic Design, Firenze (Italy).

- Arcidiacono G., Calabrese C., Rossi S. (2007) *Six Sigma: Manuale per Green Belt*, Springer-Verlag Italia, Milano.

- Breyfogle F.W., (2003) *Implementing Six sigma: smarter solutions using statistical methods*, Wiley, Hoboken.

- Imai M. (1997) *Gemba Kaizen: A Commonsense, Low-Cost Approach to Management*, McGraw-Hill, New York.

- Marchwinski C., Shook J., Schroeder A. (2008) *Lean lexicon: a graphical glossary for lean thinkers*, Cambridge, The Lean Enterprise Institute.

- Liker J. (editor) (1997) *Becoming Lean: Inside Stories of U.S. Manufacturers*, Productivity Press, Portland.

- Keats B.J., Montgomery D.C. (1996) *Statistical applications in process control*, M. Dekker, New York.

- Ohno T. (1988) *Toyota Production System: Beyond Large-Scale Production*, Productivity Press, Portland.

- Pyzdek T., Keller P. (2010) *The Six Sigma Handbook*, McGraw-Hill, New York.

- Rother M., Harris R. (2010) *Creating continuous flow*, CUOA Lean Enterprise Center, Massachusetts.

# References

- Rother M., Harris R., Wilson E. (2003) *Making Materials Flow: a lean material-handling guide for operations, production-control, and engineering professionals*, Lean Enterprise Institute, Cambridge.
- Rother M., Harris R. (2010) *Creating continuous flow*, CUOA Lean Enterprise Center, Massachusetts.
- Rother M., Shook J.R. (2003) *Learning to see: Value-Stream Mapping to Create Value and Eliminate Muda*, The Lean Enterprise Institute, Cambridge.
- Shingo S. (1986) *Zero Quality Control: Source Inspection and the Poka-Yoke System*, Productivity Press, Stamford.
- Shingo S. (1989) *A study of the Toyota Production System From an Industrial Engineering Viewpoint*, Productivity Press, Cambridge (MA).
- Womack J.P., Jones D.T. (2003) *Lean thinking: banish waste and create wealth in your corporation*, Free press, London.